Mastering
Microsoft Teams

Empowering Your Business with Advanced Features for Effective Communication, Project Management, and Team Connectivity

Elliot Mercer

Copyright © 2024 **Elliot Mercer**

All Rights Reserved

This book or parts thereof may not be reproduced in any form, stored in any retrieval system, or transmitted in any form by any means—electronic, mechanical, photocopy, recording, or otherwise—without prior written permission of the publisher, except as provided by United States of America copyright law and fair use.

Disclaimer and Terms of Use

The author and publisher of this book and the accompanying materials have used their best efforts in preparing this book. The author and publisher make no representation or warranties with respect to the accuracy, applicability, fitness, or completeness of the contents of this book. The information contained in this book is strictly for informational purposes. Therefore, if you wish to apply the ideas contained in this book, you are taking full responsibility for your actions.

Printed in the United States of America

TABLE OF CONTENTS

TABLE OF CONTENTS III

INTRODUCTION 1

OVERVIEW OF THIS BOOK 2

CHAPTER 1 3

UNDERSTANDING TEAMS 3

    DOWNLOADING AND INSTALLING MICROSOFT TEAMS 3
    KEY FEATURES OF MICROSOFT TEAMS 5
        Chat 5
        Meetings 5
        Channels 5
    INTEGRATION WITH MICROSOFT 365 5
        Collaborations 5
        App ecosystem 5
        Availability 6
    OPENING AND EXPLORING TEAMS 6
    TEAMS CALENDAR 11
    REVIEW QUESTIONS 13

CHAPTER 2 14

EXPLORING NEW FEATURES IN TEAMS 14

    COLLABORATIVE LOOP 14
    DIRECT LINKING FROM YOUR TASK LIST 16
    EASY WAY TO FIND CHANNELS OR CHATS 16
    MARK TEAM NOTIFICATIONS 17
    CUSTOMIZE A DEFAULT SET OF REACTIONS 17
    THE NEW ONEDRIVE APP 18
    THE MEET APP 18
        The Stream app 19
    BROWSER INTEGRATION 20
    LIVE CAPTIONS IMPROVEMENTS 21
    REVIEW QUESTIONS 22

CHAPTER 3 23

CREATING TEAMS AND CHANNELS 23

    Creating a team 24
    From a group 26

From scratch	26
From template	28
Joining a team	30
MANAGING TEAMS AND CHANNELS	31
Managing teams	32
CREATING TAGS	34
Channels	34
LEAVING A TEAM	36
Rejoining a team you left	37
Hiding a team	37
Deleting a team	37
REVIEW QUESTIONS	38

CHAPTER 4 39

TEAMS CHATS AND CONVERSATIONS 39

TEAMS CHATS	39
POSTING AND RECEIVING MESSAGES	41
STARTING A NEW CONVERSATION	42
ADDING FUN TO YOUR CHATS	42
FORMATTING YOUR CHATS	43
SHARING AND ACCESSING FILES VIA CHAT	44
ADDRESSING A SPECIFIC PERSON	45
BOOKMARKING, EDITING, AND DELETING POSTS	46
SENDING PRIVATE MESSAGES	48
REVIEW QUESTIONS	51

CHAPTER 5 52

SETUP MEETINGS AND CALLS 52

BEFORE YOU INITIATE VIDEO AND AUDIO CALLS	52
STARTING A MEETING	55
Initiate a call on demand	56
Schedule a meeting	63
Joining a scheduled meeting	65
SHARING YOUR SCREEN	67
Making live presentations in meetings	69
Brainstorming ideas in a meeting	72
ADVANCED MEETING SCENARIOS	77
REVIEW QUESTIONS	81

CHAPTER 6 82

USING MICROSOFT TEAMS FOR PROJECT MANAGEMENT ... 82
Structuring ... 82
Communication ... 82
Sharing files ... 83
Integration of external tools ... 83
Review Questions ... 84

CHAPTER 7 ... 85
USING MICROSOFT PLANNER IN TEAMS ... 85
My Day ... 85
My Tasks ... 87
My Plans ... 88
My Teams ... 91
Important settings ... 92
Old vs. new planner ... 92
Review Questions ... 94

CHAPTER 8 ... 95
MICROSOFT TEAMS FOR EDUCATION: EXPLORING NEW FEATURES ... 95
Annotate a PDF ... 95
Reminders ... 98
Extend due date ... 98
Bulk feedback ... 99
Reflect mindful coloring book ... 100
Reflect for staff ... 101
Noise suppression for reading progress ... 102
Warning for missing attachment ... 103
Updated editing option for students ... 104
Use of tables ... 105
Updated Turn in celebrations ... 105
School connection app ... 106
Review Questions ... 108

CHAPTER 9 ... 109
TIPS AND TRICKS ... 109
The basics ... 109
Mute unrelated conversation threads ... 110
Create a task from a post ... 110
The activity center ... 110

  How to set your pop-up windows *112*
 SCHEDULE AN OUT-OF-OFFICE NOTIFICATION 113
  Adding new tabs *113*
  Using the search bar *115*
 REVIEW QUESTIONS 118

CHAPTER 10 119

TROUBLESHOOTING TEAMS 119

 TEAMS NOT WORKING? 119
 SPELL CHECK NOT WORKING? 120
 REVIEW QUESTIONS 123

CHAPTER 11 124

TEAMS APP AND ADD-ONS INTEGRATIONS 124

 APP INTEGRATIONS 124
  Meet app *124*
  OneDrive app *125*
  Planner *126*
  Viva insights *126*
  SharePoint *127*
 REVIEW QUESTIONS 128

INDEX 129

INTRODUCTION

If you are new to Microsoft Teams and you need a little bit more information on how to use it, let's say your employer just switched over to using it and you don't know what is going on or how to use it at all, this is the guide for you. Imagine being able to communicate more easily, manage projects better, and get more done with your team – that's what Microsoft Teams 2024 can help you achieve. Staying up-to-date with the latest tools can make a big difference in how successful you are and this book will help you understand how to use Microsoft Teams in 2024, covering all the new features and tips to work better with your team. With Teams, you can make teamwork smoother, communication clearer, and work more efficiently, and with this book, you'll learn helpful information and tools to use Microsoft Teams effectively. So whether you're new to the platform or already familiar with it, there's something here for you to improve your work with others. Why do you need this book? Well, learning how to use Microsoft Teams effectively can make your teamwork smoother and help you get more done. To get the most out of this book, take your time going through each chapter, trying out the activities, and practicing them in your everyday work. Before you know it, you'll be a pro at using Microsoft Teams to make work easier and more productive for your team.

OVERVIEW OF THIS BOOK

This book is about the newest features and improvements in Microsoft Teams, a popular collaboration platform. It is designed to be a complete guide, teaching you how to use all the different parts of Teams with ease.

CHAPTER 1 starts by showing you how to download, install, and explore the key features of Microsoft Teams, including the Teams Calendar for managing schedules and collaborating.

In CHAPTER 2, you will learn about the latest features in Microsoft Teams, such as the Collaborative Loop, direct linking from task lists, better search, and improvements to live captions and browser integration. These new features can make workflows and team collaboration even smoother.

CHAPTER 3 explains how to create and manage Teams and Channels, which are important for effective communication and organization. You will learn how to set up Teams from various sources, manage team members, create tags, and customize channel settings.

Effective communication is key in Microsoft Teams, and chapter 4 talks about Teams Chats. You will learn how to post and format messages, share files, and use advanced features like addressing specific people and bookmarking important conversations.

CHAPTER 5 will guide you through the process of setting up virtual meetings and calls, from initiating on-demand calls to scheduling and joining meetings. It also shows you how to use advanced features like screen sharing, live presentations, and brainstorming sessions.

Using Microsoft Teams for project management can greatly improve collaboration and productivity. Chapter 6 shows you how to structure projects, facilitate communication, share files, and use external tools within Teams.

In chapter 7 you will discover the power of Microsoft Planner, a task management tool that is part of Microsoft Teams. This chapter shows you the different components of Planner and how it has evolved over time.

Chapter 8 focuses on the latest features and tools in Microsoft Teams specifically for educators and students, such as annotation, reminders, feedback, and noise suppression.

Chapter 9 provides a collection of useful tips and tricks to help readers get the most out of Microsoft Teams, from muting irrelevant conversations to creating tasks from posts and customizing pop-up windows.

Addressing common issues is very important, and chapter 10 guides you through troubleshooting steps for resolving problems, such as Teams not working or spell check not functioning.

Chapter 11 talks about the various apps and add-ons that you can integrate with Microsoft Teams, including the Meet app, OneDrive, Planner, Viva Insights, and SharePoint, showing you how to leverage these integrations.

Are you ready to learn about Microsoft Teams? Let's get started.

CHAPTER 1
UNDERSTANDING TEAMS

Microsoft Teams is a collaboration platform developed by Microsoft. It provides a centralized hub for teamwork, allowing users to chat, make audio and video calls, share files, and collaborate on various projects and tasks. Teams is designed to facilitate communication and collaboration within organizations, whether those are small businesses, educational institutions, or large enterprises.

Downloading and Installing Microsoft Teams

Search Microsoft Teams download in your web browser, click on "Download Microsoft Teams Desktop, then click on "Download app for desktop."

Next, click on "Download Teams for home" or you can download Teams for work or school. Wait for Microsoft Teams to download.

After downloading, click on the Open button to start installing Microsoft Teams. It will take about a minute to install.

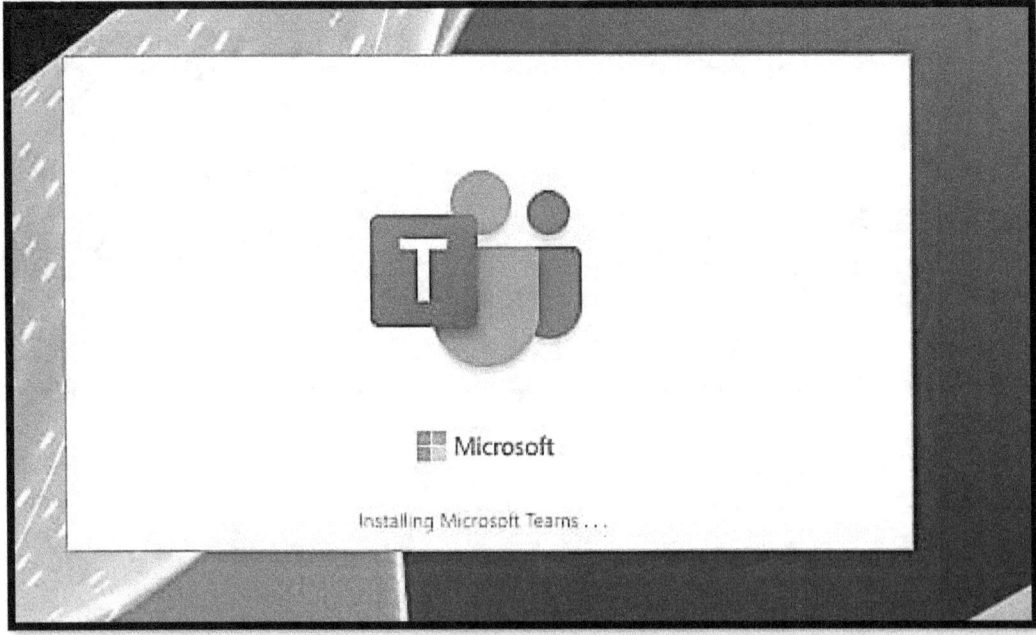

After installing, sign in with your Microsoft account. Microsoft Teams will open. You can now use Teams to join a meeting or start a conversation.

Key features of Microsoft Teams

So what are some of the key features of Microsoft Teams?

Chat

Users can have one-on-one or group chats with colleagues, share files, and use emojis, stickers, and gifs to express themselves.

Meetings

Teams offer audio and video conferencing capabilities, allowing users to schedule and join meetings with colleagues or external participants. It supports screen sharing, virtual backgrounds, and recording options.

Channels

Teams organize conversations and files into channels that can be dedicated to specific projects, departments, or topics. Channels provide a structured way to collaborate and keep discussions organized.

Integration with Microsoft 365

Teams integrates with other Microsoft Office applications such as Word, Excel, and PowerPoint, enabling users to create, edit, and share files within the platform.

Collaborations

Multiple team members can collaborate simultaneously on documents. Teams seamlessly integrate with other Microsoft 365 services like SharePoint, OneDrive, and Outlook. This integration allows for easy access to files, calendars, and other resources within the Teams' interface.

App ecosystem

Microsoft Teams supports a wide range of third-party Integrations and apps, allowing users to enhance their experience and connect with other tools they use in their workflows.

Availability

Microsoft Teams is available as a desktop application for Windows users and Mac users. It also has a web-based version accessible through browsers and mobile apps for IOS and Android devices. It has gained significant popularity especially since the pandemic as remote work and virtual collaboration have become increasingly prevalent.

Opening and Exploring Teams

Now we'll open up Microsoft Teams from the portal, take a look at some of the basics, and explore the interface. We're going to look at our app launcher and you should find it here as Microsoft Teams. Now it's worth noting that we're accessing through the web browser. You can also download the desktop application and access Teams that way. The interface is the same. Now the first thing you'll notice when you go into Teams is that the main action takes place in this left-hand menu that we have running vertically down the left-hand side. These are all of the different parts of Teams.

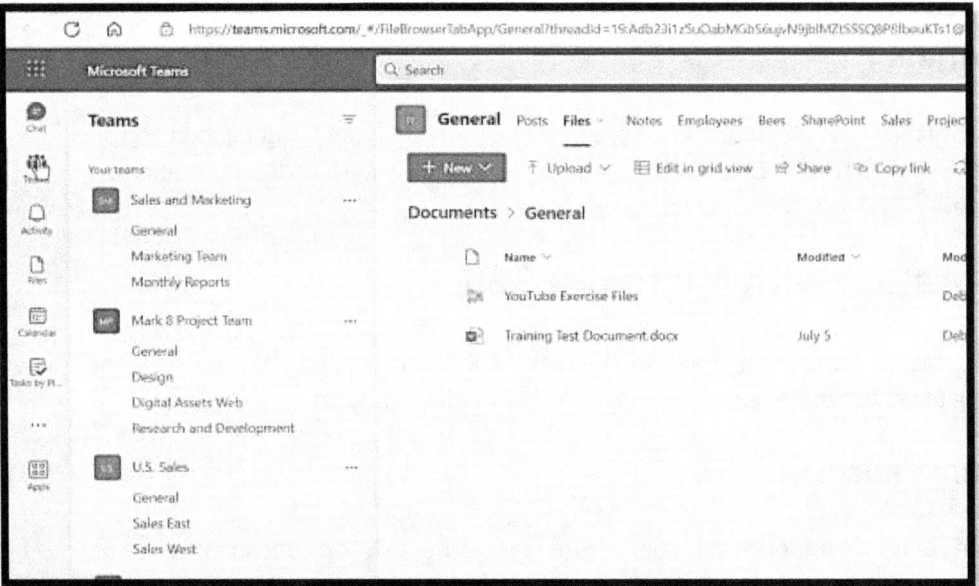

The one highlighted in purple is the one that you're currently clicked on and by default, it will select Teams so when we click on Teams we can see a big long list of all of the teams that we have access to. For example, Sales and Marketing is a team. These little folders that you see underneath them are referred to as Team channels, and channels are there to direct specific conversations. So, within the Sales and Marketing team, we have a channel specifically for the Marketing team. This is where all of the marketing-type posts, messages, and files would be shared. We have another channel for monthly reports and that is where all discussion related to that topic takes place. You can have as many channels as you like under here to get very specific about certain topics within a team. Now one thing you will notice is that every team that's created has a general Channel by default. This is where all general conversations will take place and if we click on one of the teams down here, you can see we've got a conversation going on here. That brings us to our next point: if you take a look at the general Channel (this is the same for every channel that we have) you can see at the top here we have different tabs. This is more like what you'll see when you first create a new channel. You'll just have two tabs at the top for posts and files. In the Post section, this is where all conversation takes place so if you want to have a chat with your team members this is where you would do it and you can see right at the bottom we have a new conversation button. The Files tab is where we can see any files that have been shared within this team channel so it just gives us easy access to things that we're currently working on. Next to that, we have a plus symbol which is going to allow us to add our tabs and we can connect to all different kinds of applications from here. We will go into this in more detail a bit later on but for now, the main takeaway here is that we have teams and channels and then within each of the channels we have different tabs running across the top.

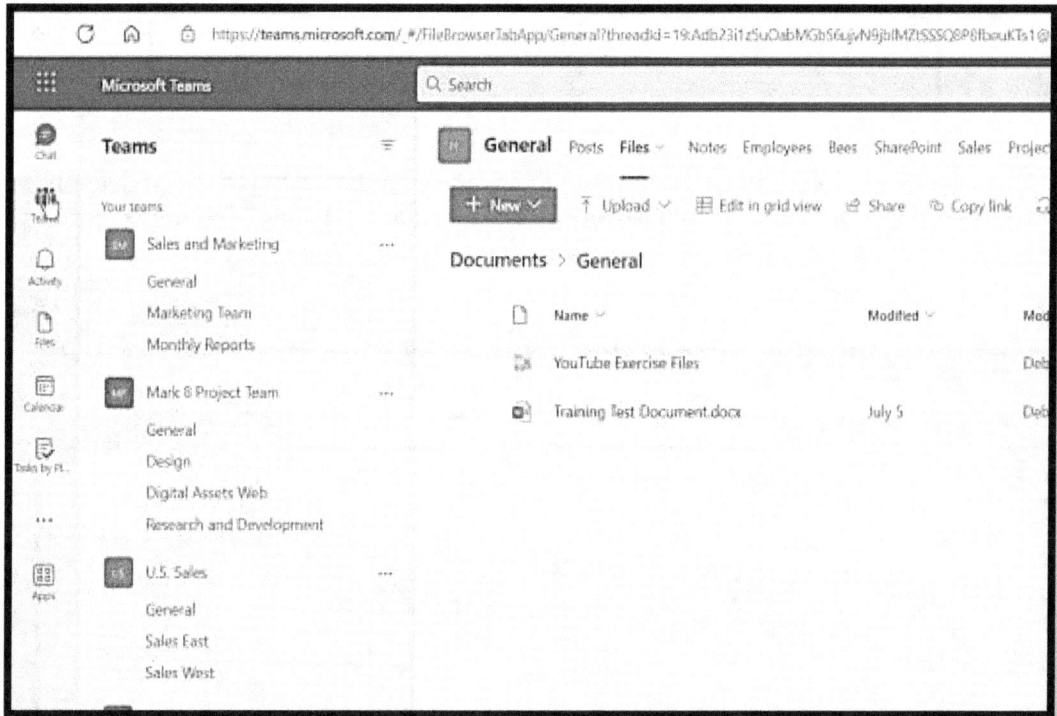

Another thing you'll also notice is that next to each team and each channel, if you hover your mouse over it you get three dots at the side here and if you click this it's going to open up a little contextual menu that just gives you more options for executing actions and managing your teams and your channels so just be aware of that because we'll be going in and out of this as we go through this section.

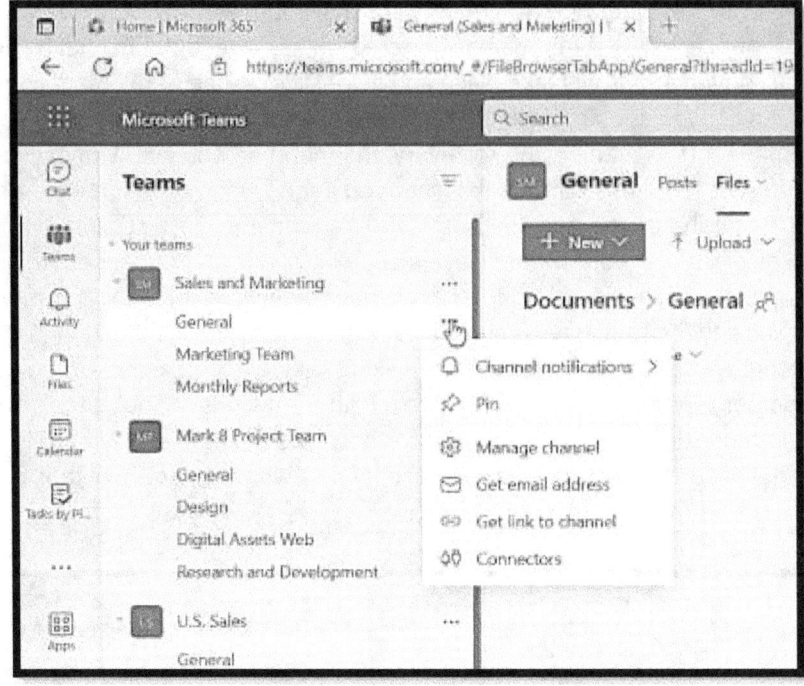

The icon just above is **Chat**. This is where we go to have private chats with people. It might be that you don't necessarily want to message everybody within a team's Channel, you might just want to have a private conversation with one member from that channel and this is where you can go to have private chats.

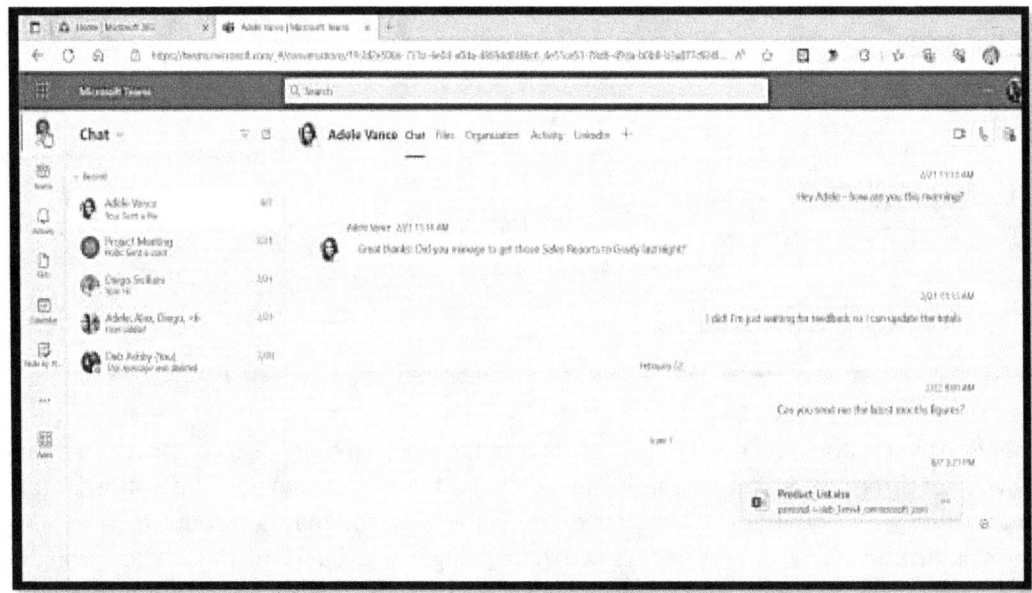

The **Activity** option here is a feed so what you're going to see here are all of the updates that are going on across your teams. This is particularly useful if you have people @ mentioning you because you can come here and see immediately who's mentioned you and which team and channel you need to go to to reply to that mention.

Underneath that, we have a **Files** area. This is just going to take us to the team's files library and this is going to show us all of the files that have been shared within the teams we have access to but it also allows us to filter by different file types as well. If we just want to see all the Word files we can do that and we can also upload documents from our PC straight into this file library. We can create brand-new documents, workbooks, and presentations by clicking the "New" drop-down. Again, we're going to discuss all of this in much more detail a bit later on.

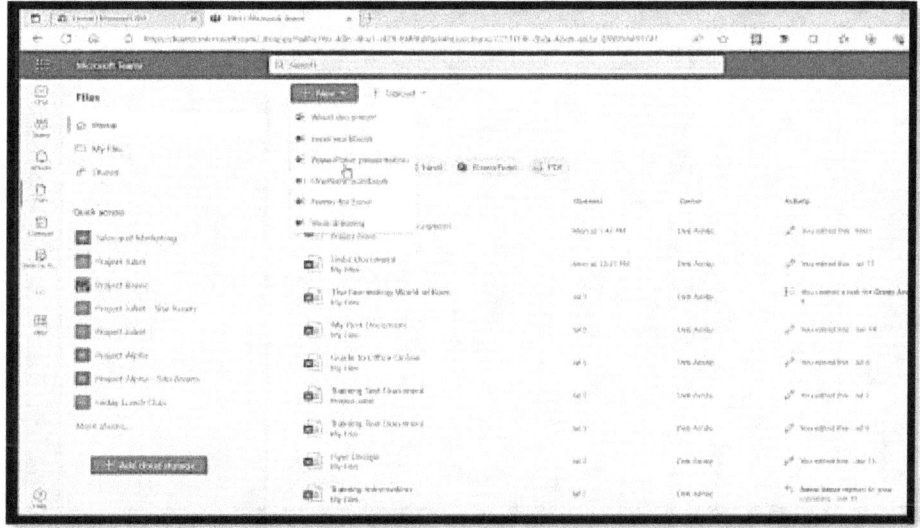

We also have access to a team's **calendar**. This is great if you want to have a shared calendar with your team.

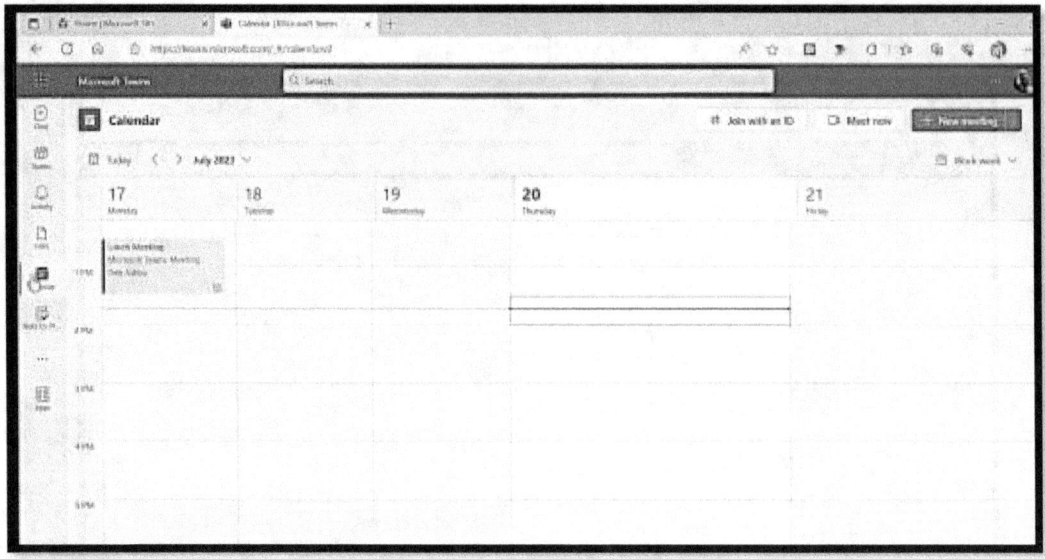

Then underneath that, we have something called **Tasks by Planner**. This is another application that we've added to keep track of all of our team tasks and this can be added using the Apps button at the bottom. If we click Apps it gives us access to all of the third-party apps that we can add to teams to extend our experience and make it more functional.

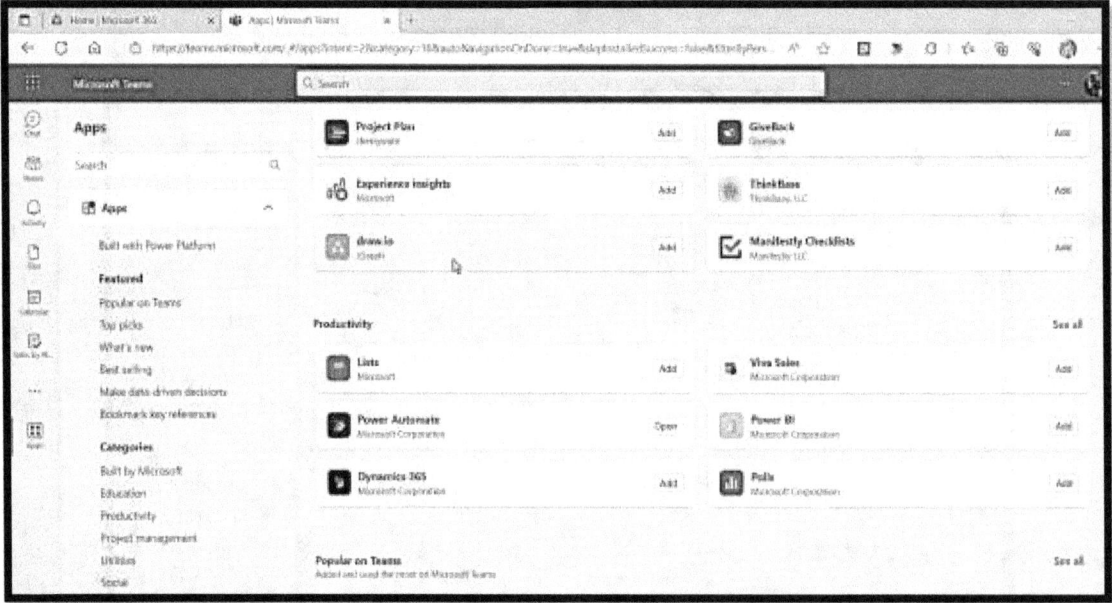

The final thing to mention is right at the top we have a **search bar**. This is a universal search that allows you to search for all manner of things across your different teams and there are a few little tricks and keyboard shortcuts that we can use here which makes finding things a lot simpler. We'll cover this too in more detail later.

Teams Calendar

We've got the calendar as well and it's straightforward. What you've got up here is an option to join with an ID, Meet now, and New meetings.

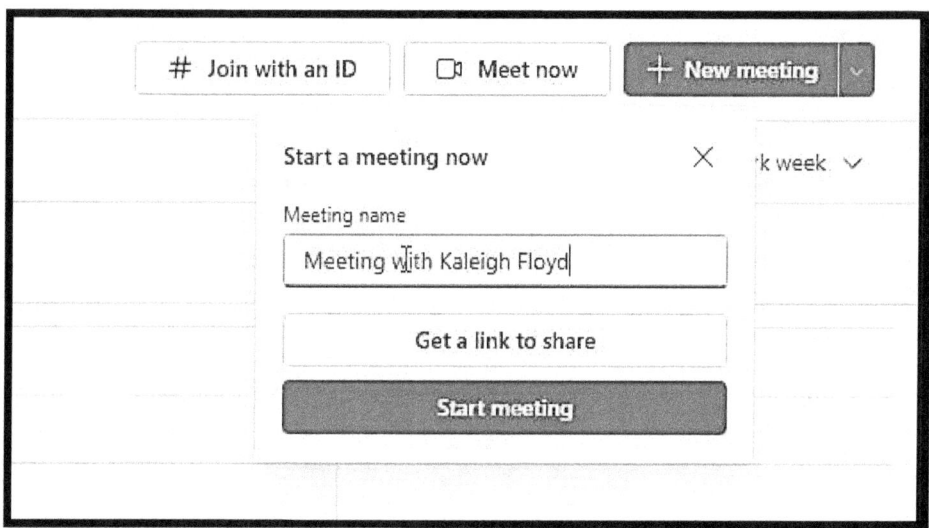

When you join with an ID you can choose to join somebody's meeting if they've given you a specific ID. When you want to Meet Now, what it will do is create a quick meeting at that very moment and you can get the link right here and send it to somebody. That's if you wanted a quick and impromptu meeting at that moment. Then you've got the New meeting schedule and when you click on that, it is going to pull up a window where you can choose to add a title, add required attendees, then you can indicate if you want people to decide whether they want to come to the meeting or not. Next, you're going to say the date that you want the meeting to be set and the time. You can also choose if the meeting repeats or not and if it's a daily, weekly, or monthly meeting, and if you want to connect it to a channel you also have the opportunity to do that as well. If you do connect it to a channel, you can send personal invites to everybody who's on that team. Another thing you can do is choose a location for the meeting or you can choose if it is an online meeting or not. When you have an online meeting you have the option to have this Lobby where people can sit and wait for the meeting. Now you can choose to click down on the last button here and allow only organizers and co-organizers to bypass the lobby. When you do this, the organizers and the co-organizers have the chance to be in the meeting and talk before letting the other people into the meeting.

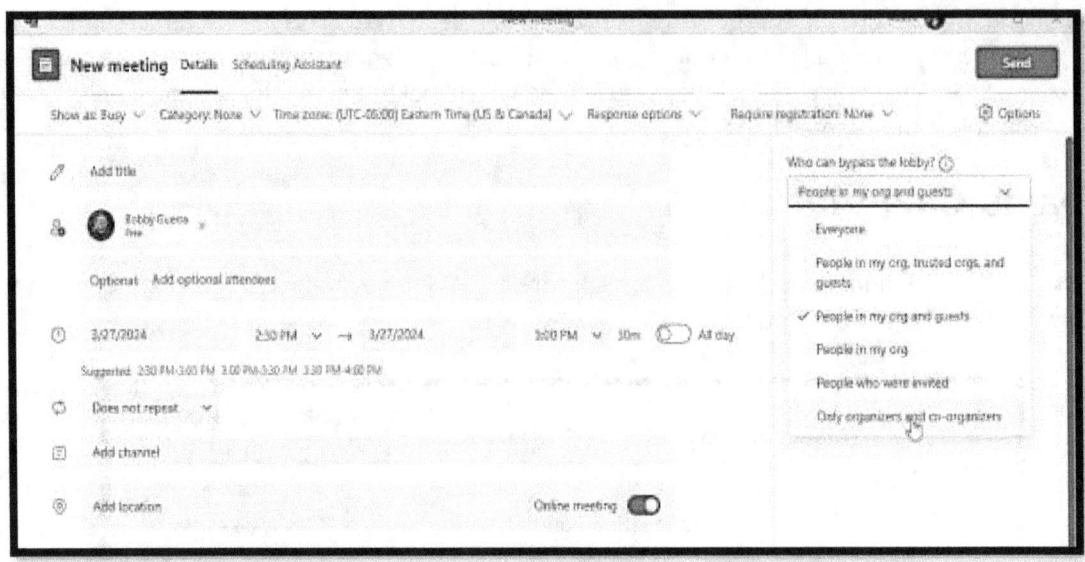

Next, you've got the details of the meeting and then you also have the option for an agenda. Once you're done with all this information you can choose to send the meeting invites which it will send immediately to them and you'll see it here on your calendar.

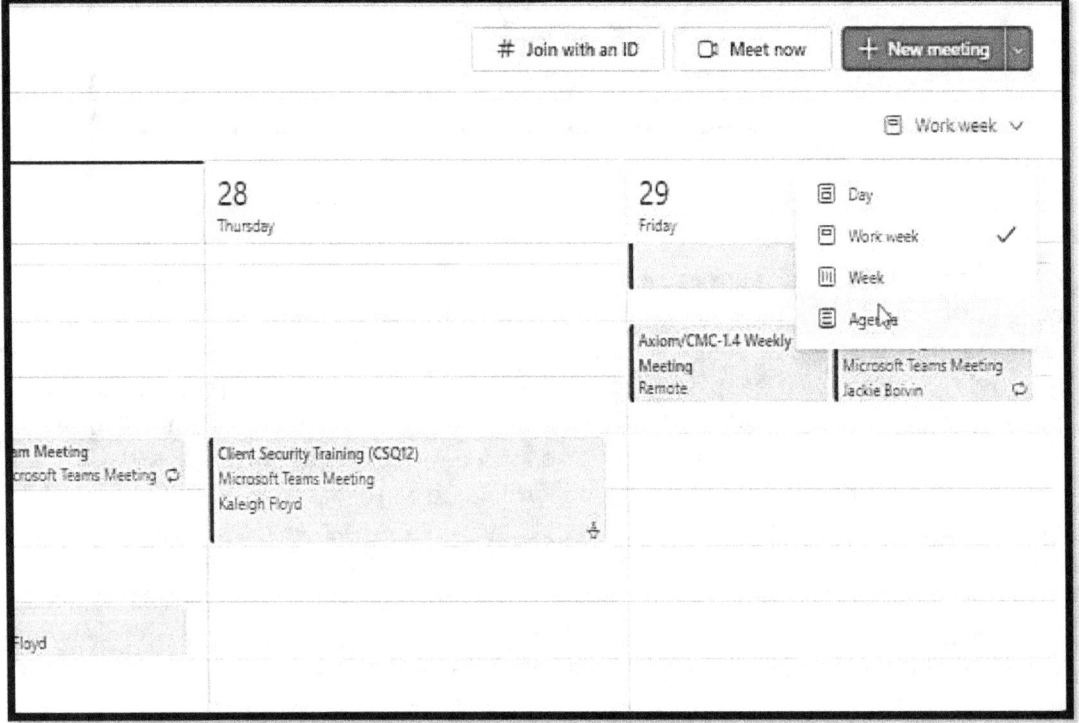

There are a few ways that you can view your calendar: you can choose to view it based on the Work week, view by day, or week, or use the new view called Agenda which only shows where you have meetings. If you use the last option, you'll notice that it doesn't show the days you don't have any meetings because there's nothing on your agenda. You'll want this view especially when you're trying to find meetings pretty quickly.

Review Questions

1. Download and install the Microsoft Teams desktop app on your computer.
2. Customize the layout of the Teams app to suit your preferences.
3. Open the "Calendar" section of the Microsoft Teams app and explore the different calendar views and settings available in the Teams app.

CHAPTER 2
EXPLORING NEW FEATURES IN TEAMS

In this chapter, we'll be showing you the new features in Microsoft Teams. This includes Loop components in channels, new apps like Stream, Meet, and OneDrive, the ability to customize your own reaction tray, and a whole lot more.

Collaborative Loop

Components are now supported in channels and not just chat. At the bottom, click "Start a post" and you'll get a subject. Right here, you'll click the plus button - this is the consolidated way that they've done the UI in the new Teams and you can add "Collaborate with Loop."

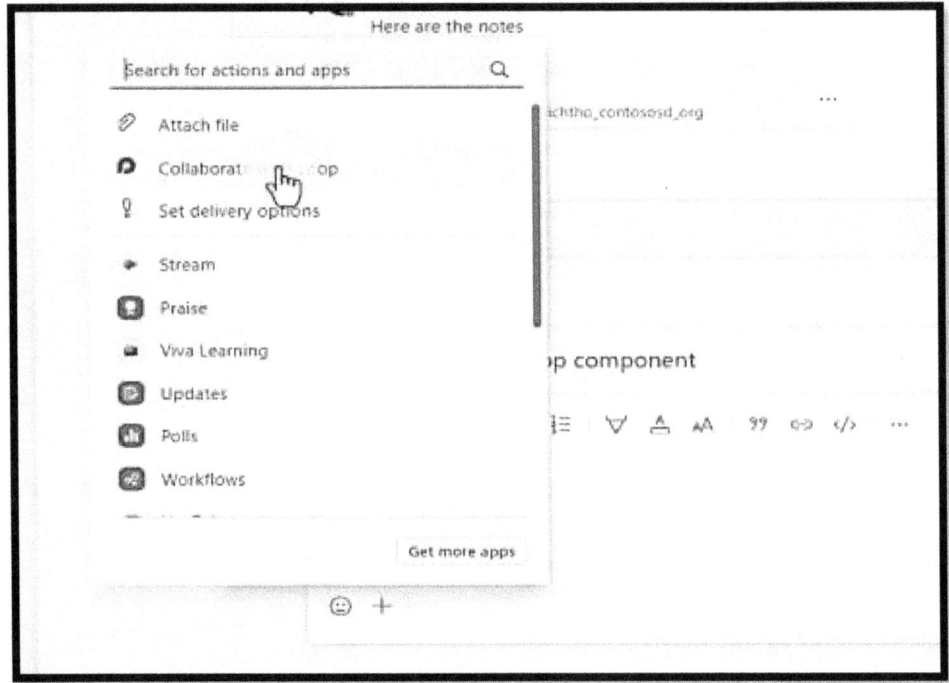

Here, you've got all these different options: list checklist, Q&A, tables, task list and these are all collaborative just like the way Loop works in all the other apps whether it is Whiteboard, Outlook, or Word. You can choose a task list, for example, to make your first component. Give it a title then start adding tasks to this. You can add a new task here and do a quick post of this message and you'll see your task planning Loop component is in the main Channel now others can go and add things to this as well, so it's easy to do collaborative Loop components.

This is just an example of what you can do; there are many other loop components you can do. You can copy components by just clicking "Copy Component" to single click and put that in the email, you can see who has shared access and you can also look at shared locations. There are a lot of fun things to do with Loop, just note that it is now supported in channels inside of Teams.

Direct linking from your task list

This second new feature is also from Loop and that is you can now see the direct linking from your task list components to planner or to-do.

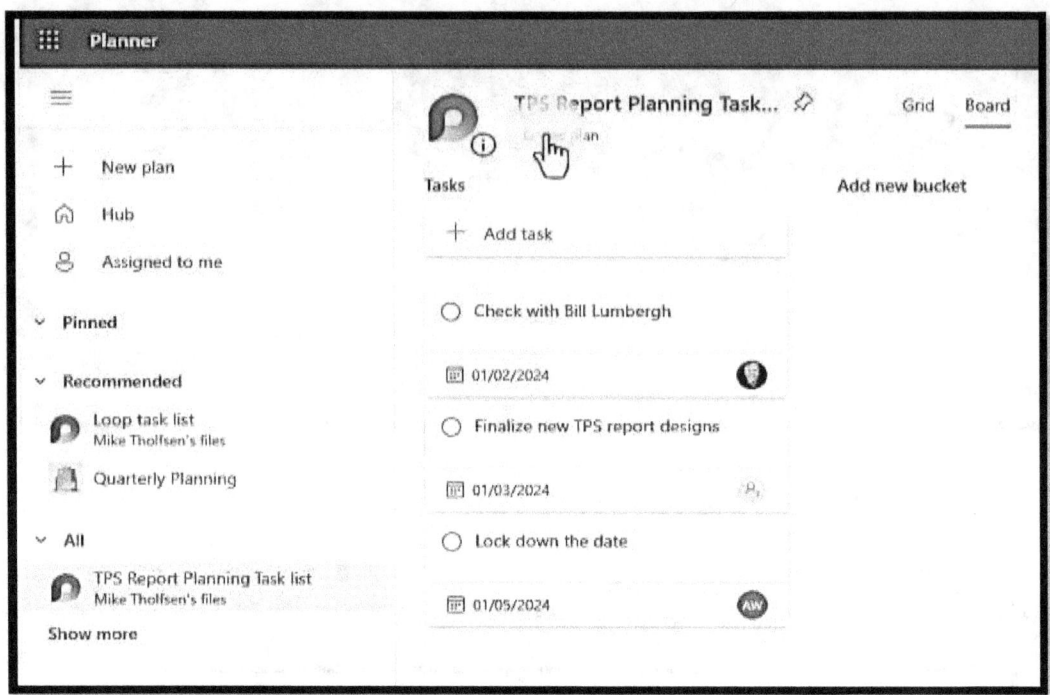

Right here it has a Tasks app. If you click this down you can say open in Planner and you'll see there's a Planner board that was automatically created by that Loop component. What's also nice is that this automatically puts the task into to-do. If you open up the to-do list you can see the date it was automatically popped right into the to-do list which is a free app in Microsoft 365. What this does is help you find something in a Channel or a chat.

Easy way to find channels or chats

This is super easy to use. Let's say you're in a channel, if you do CTRL+F that open up the pane on the right to find in a channel. If you type in something, you get a lot of instances and if you click Enter you'll see how it pulls up quickly then you can click to go right to that message and find whatever you want. This works the same in a chat. Go to a chat right here, do CTRL+F, search for a report, and hit Enter. You'll see lots of entries in this chat all about the report. You can now go and click around and find the exact instance that you're looking for.

Mark Team notifications

The fourth new feature lets you mark all team notifications for a specific team as read with a single click. Let's say in the Product team we've got three different notifications right here at the top. We're going to hit that three-dot menu then choose "Mark all as read" and immediately all three of those notifications are marked as read, and they'll go away.

Customize a default set of reactions

If you hover over a chat, you'll get a little set of reactions. If you click the smiley face with a plus you're going to have this reaction set. Now right here there's another little smiley face that says "Customize default reactions." If you click on that you'll get four as defaults but then you can turn off some of these and add yours then hit Save. Now when you hover over a message those are the default reactions that'll pop up.

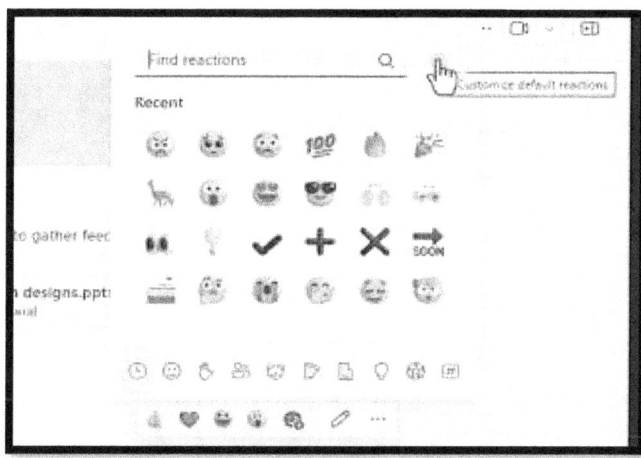

The new OneDrive app

The Files app over on the left-hand rail is now the OneDrive app and it pulls in the entire new OneDrive. Click on that and it's going to load up your OneDrive which is the same OneDrive you're going to see in the web version of OneDrive. Using the Upload button, you can add new files and upload them. You have all the filtering right here including Word, Excel, PowerPoint, and others. You can browse by people, you can browse by meetings, and all of your quick access to other teams is right down here. This gives you quick access to other files and all of the benefits of the brand-new OneDrive are now here in Teams.

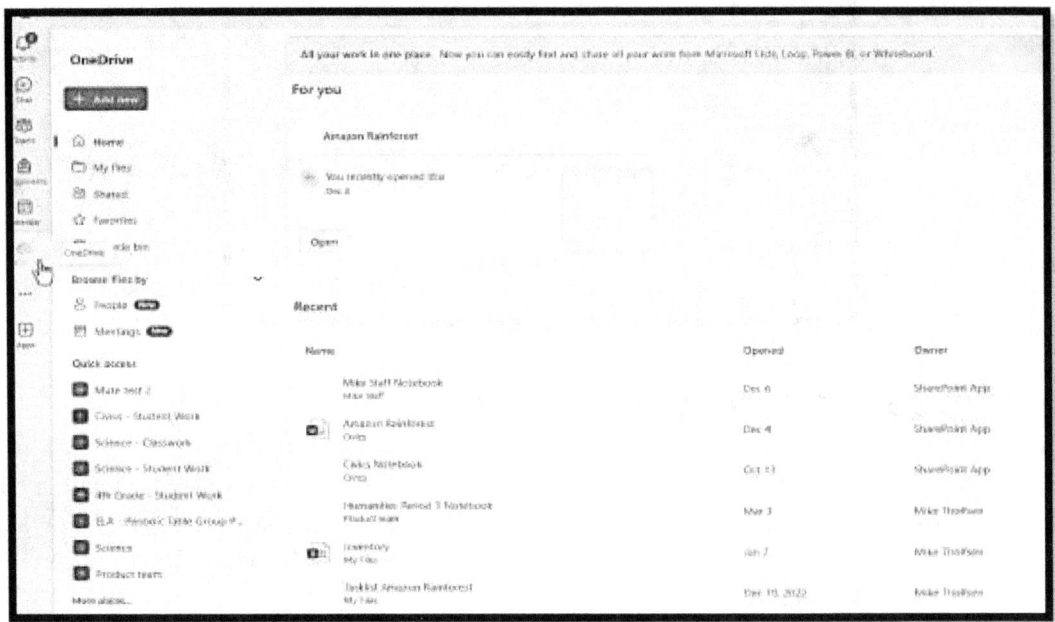

Note that when you're in a team you will still have the classic files up at the top. In the general channel, you can still access the files so if you're a teacher using class teams and you're wondering about class materials, that's all the same - you'll find this unified OneDrive that pulls all of your files together into one place on the left side.

The Meet app

This has been added to the new Teams. If you click on Meet you can get this app, which is the meetings app. You can right-click and pin this so it doesn't go away. You have all your meetings inside here, including the ones that you've had or the ones coming up. You've got recent meetings also down here with their recordings and if you want to view the recap you can go and click on that option to view that recap.

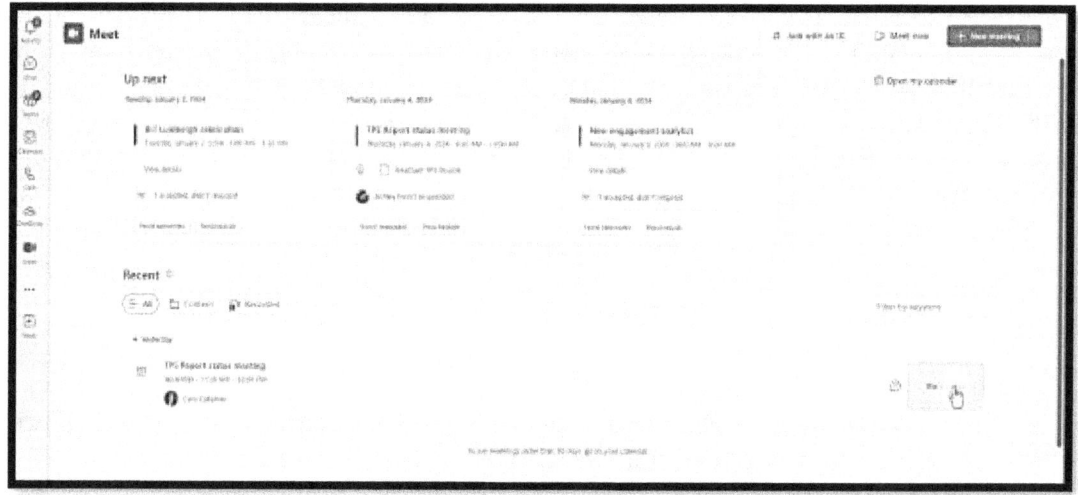

What's also nice in the Meet app is you have things like sending reminders or rescheduling. Let's say some of the people haven't responded, when you click the "Send Reminder" option it pulls up a chat window, and that way, you can send a reminder about the meeting so it's just a nice timesaver that you can pop out and send people a reminder to respond. If you want to reschedule you can also click on that option right there and this will bring up the meeting and allow you to reschedule. The Meet app is now available for anyone just to add and like we showed before, you can pin it quickly by right-clicking.

The Stream app

Microsoft Stream is our video tool for inside of Enterprises and school organizations and Stream has a brand new homepage in the main web area. To pull this here into Teams you're going to right-click and pin that.

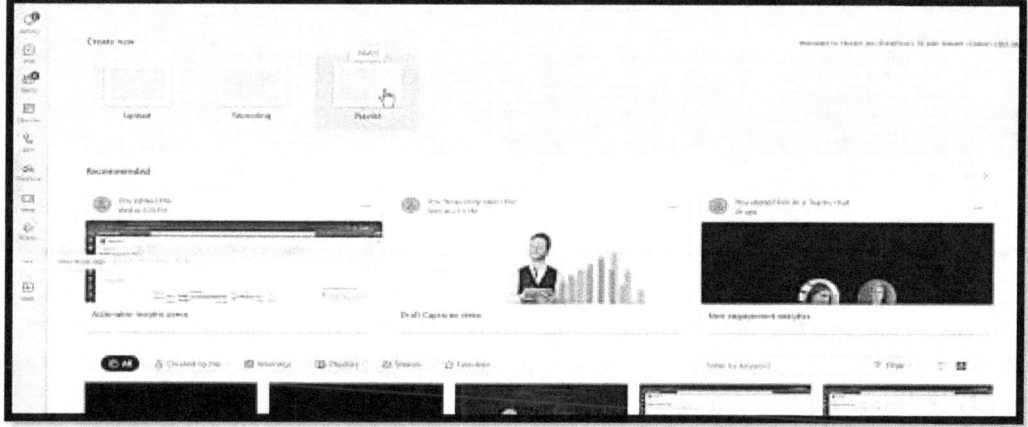

You can do things like upload videos, do screen recordings, and create playlists. It pulls together all of your recommended videos and all of your videos right here. You can go to the Grid view or put it into the list view. You can see things created by you, check for meetings that were recorded, see playlists if you have any playlists, and check shared files or favorites so you can access all of Stream right here. If you go through Microsoft 365 on the homepage you can see it looks very similar; it's the same set of user interfaces and features, it's just now in Teams instead. You can also right-click to pin this just like you've done before. if you don't want to pin it you can just say unpin and make it go away.

Browser integration

This new feature automatically integrates the Edge Browser with Teams chat. When you launch a link, for example, opens up everything from the Wikipedia site right here but on the right it pulls in the context of the chat.

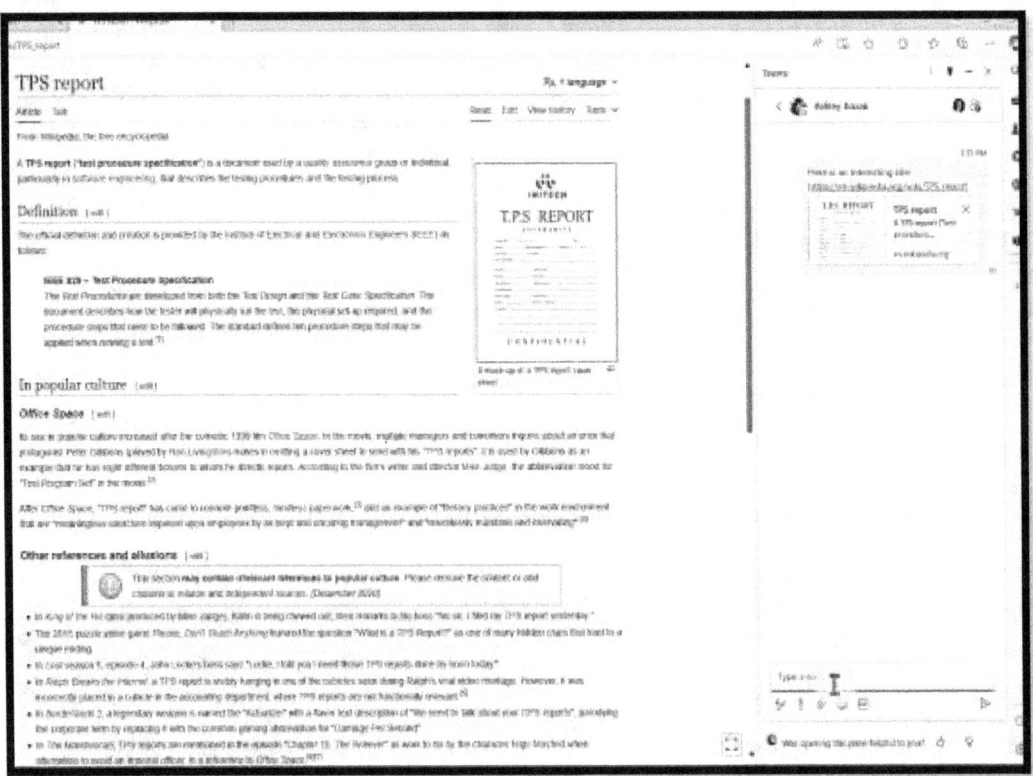

You can see a chat you had with someone about that, there's an interesting site so now you can chat back and forth with the person while you have that site open. It's a handy way to integrate your chats or if you're in channels it'll launch the same thing where it puts the chat and the context right next to the website.

Live captions improvements

This is about improvements to live captions and the way you can style and position them. Let's say you're in a meeting and you turn on live captions. Click the three-dot menu, go to Language and Speech, and then turn on Live Captions. Choose your preferred language and you can see at the bottom the live captions are rolling.

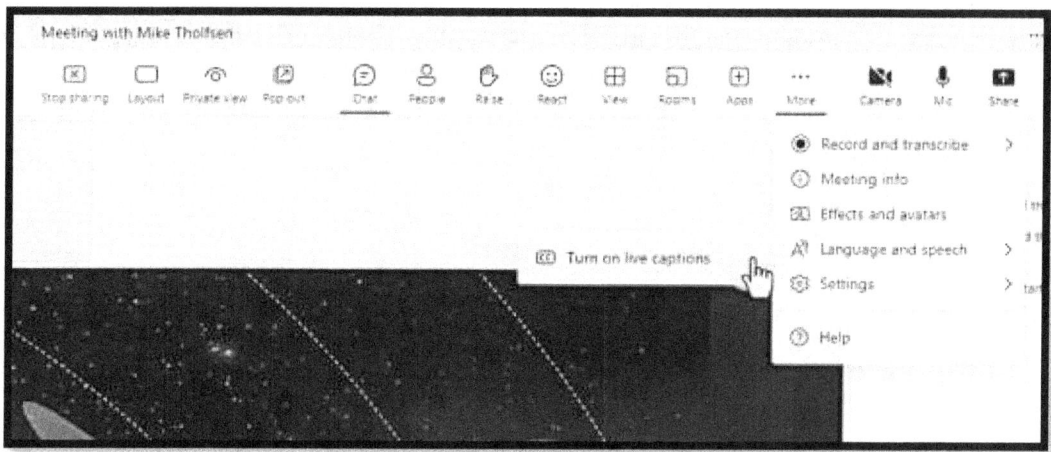

Over on the right, there's a bunch of new caption settings. Click Settings here and now you can choose different colors, height, position, and font size and change the language if you want, and as you make changes you get a preview of what it's going to look like.

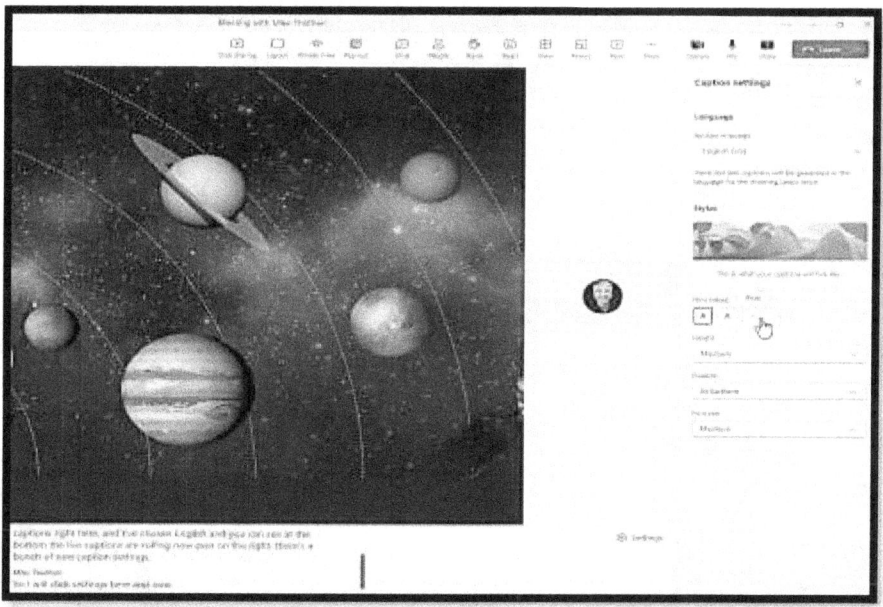

You can change the height of everything. You can make it a little bit longer or also make it smaller or medium, positioning at the bottom or the top if you want your captions up at the top so now the captions are going right across the top. Lastly, you can also change the font size. You've got lots of different options you can choose for your live captions.

Review Questions

1. Create a new Collaborative Loop and invite your team members to contribute.
2. Experiment with the different features of Collaborative Loop, such as task assignment and commenting.
3. Observe how Collaborative Loop integrates with other Microsoft 365 applications.

CHAPTER 3
CREATING TEAMS AND CHANNELS

Teams are very interesting because you've got the different channels inside of your different teams. If you go under a team you can see you've got a general Channel which is required for every single team under Microsoft Teams then you can choose to create whatever channels you'd like underneath and that's going to be different depending on what your team does and how you want to sort your information. One thing you should know is if you click on the three dots at the top here, you can choose to manage your teams, see the pending invites you have, and show the analytics of your teams.

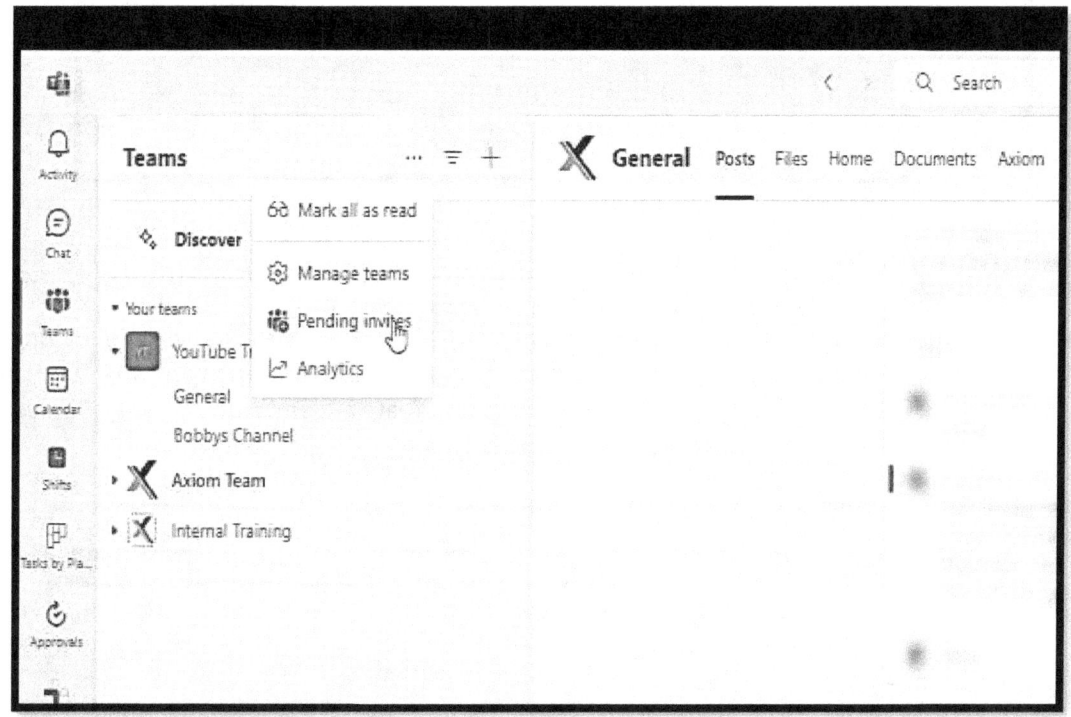

There are two types of teams that you can create in Microsoft Teams: you can create a public team or you can create a private team. The difference between the two is that if you create a private team, then as the owner of that team you need to invite people to the team for them to join. Also, the private teams are not discoverable whereas a public team is available for anybody within the organization to simply join without any approval. We're now going to take a look at an example of both. Clicking on Teams we can see all of our teams listed on the left-hand side and then right at the bottom, we have a "Join or create a team" button.

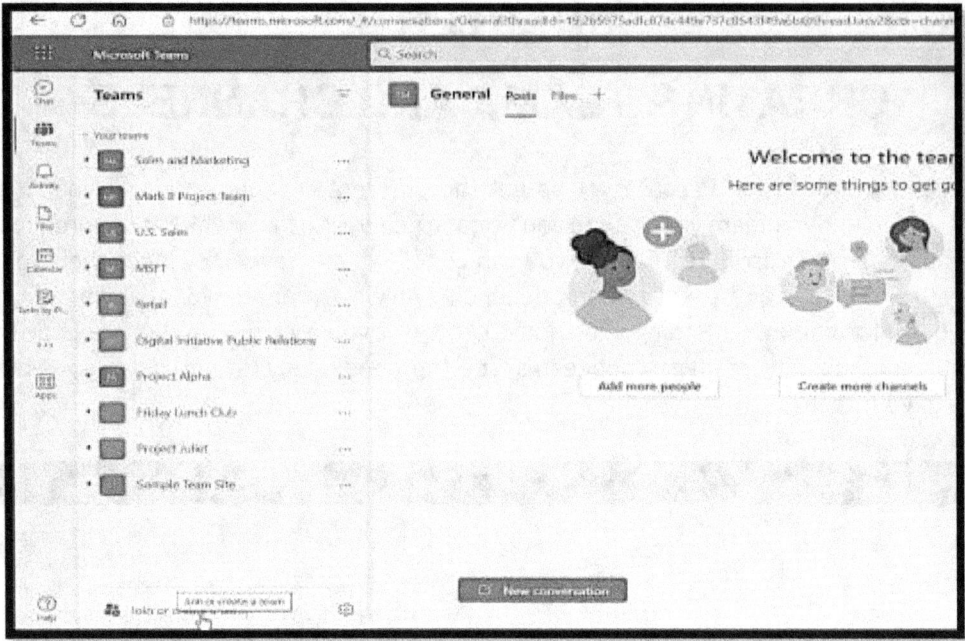

Creating a team

If we click on this it's going to take us through to a page where we have two options. The option on the right says to join a team with a code. We're going to cover this a bit later on but what we want to focus on here is how to create a team so we'll click the "Create a team" button and we then get a few different choices as to how we want to create this team.

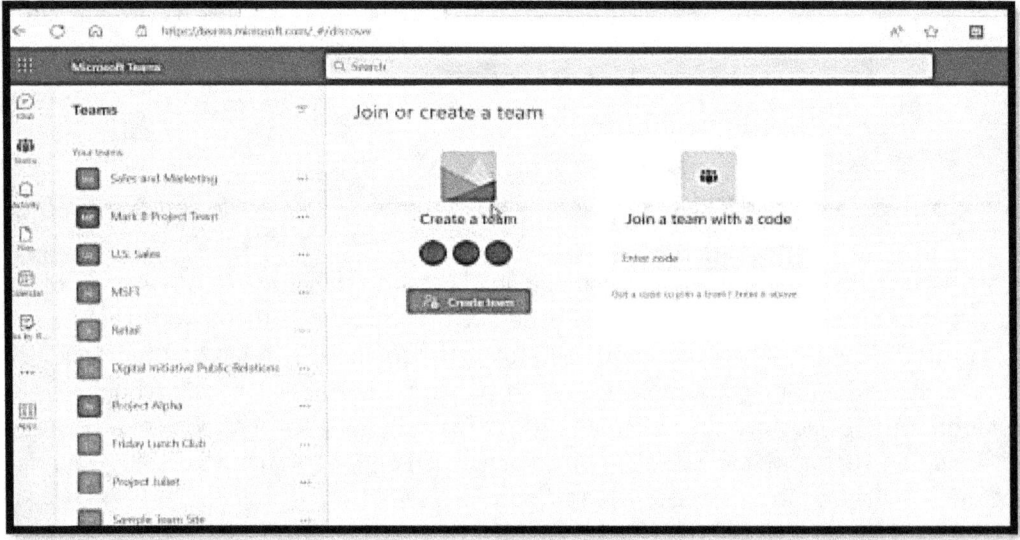

We can create a basic team from scratch and this just gives us the team structure but when we create from scratch it means that we have to add all of the members to the team ourselves.

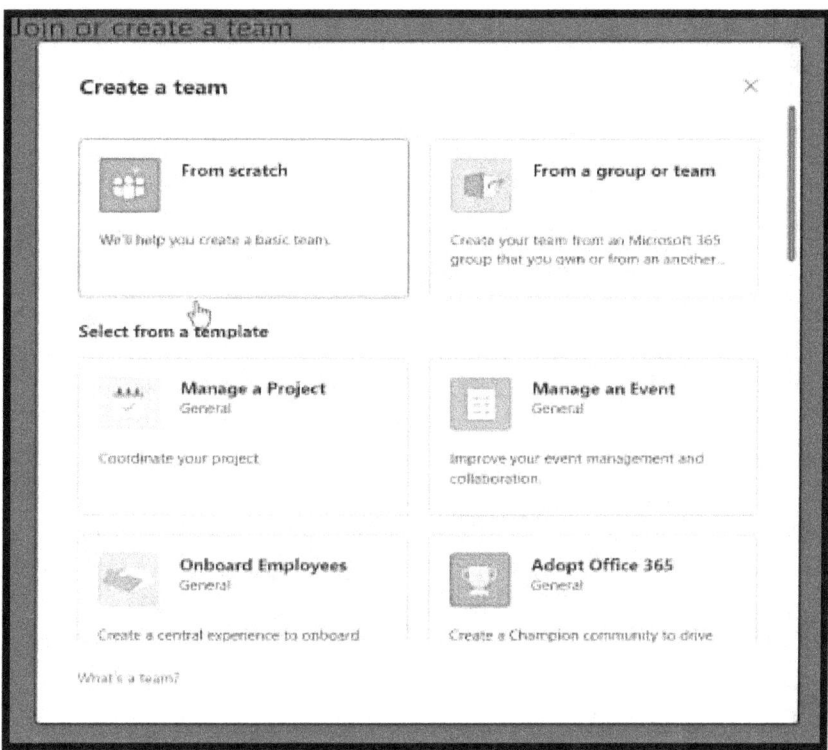

The other option that we have up here is to create a team based on an existing Microsoft Teams team or an existing Microsoft 365 group so if you already have a group set up in Outlook, with let's say 10 members you could save yourself a bit of time by creating a team based on that group and it will bring all of those members across.

The third option we have is to create a team based on a template and if we scroll through you can see that we have all different templates here. These templates define the types of channels that you're going to get; they're going to be related to the subject matter of the actual team so we'll leave it up to you to determine which one is best going to suit you but let's take a look at a couple of examples here.

From a group

For this illustration, we'll create a team based on a Microsoft 365 group so we're going to click from a group or team, choose Microsoft 365 group and we can see there it's showing us the groups that we have set up in Outlook. We'll choose one of these groups, click on "Create" and that is pretty much as simple as it gets. Now if we take a look down in the teams list there is the new team at the bottom and we have our default General channel so we can start having conversations with team members. If we click on the three dots next to the team and go into Manage team, you'll see there are all of the members, guests, and owners of the Microsoft 365 group. As the owners, we can see all the members, and of course, if we want to make somebody a co-owner we can simply bump up their access to "Own" to move them up to the top half of the screen. Creating a team in this way does save you a little bit of time if you already have a Microsoft 365 group set up and you just want to copy across the settings and the members.

From scratch

Let's take a look at another way you can create a team. We'll go back into the "Join or Create" button a team and create a team again, this time we're going to create one from scratch and this is where we get to choose a private or a public team or an organization-wide team.

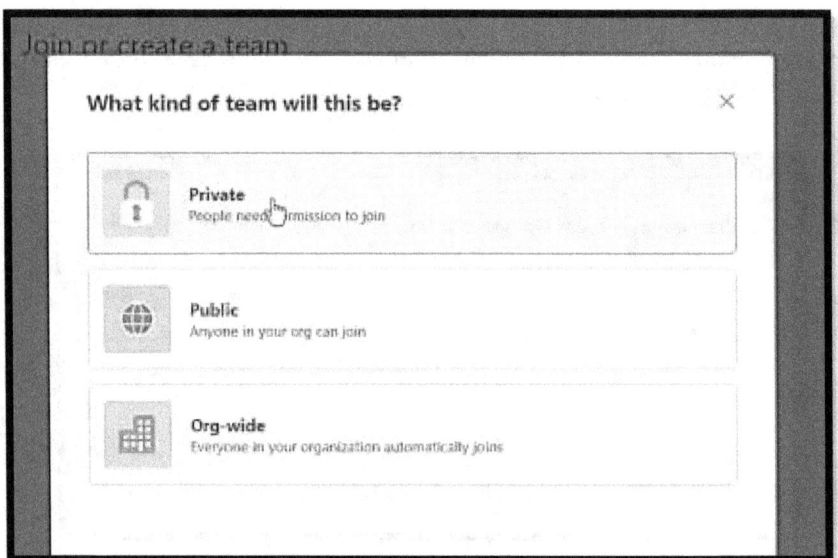

The difference between these as we said is:

- For a private group, people need permission to join.
- For a public group, anyone in your organization can join by clicking the join button.
- An organization-wide group means that everybody is automatically part of that team so as soon as you create it everybody's going to see it in their team's list within your organization.
- With a public team, they need to physically click the join button and with a private team, they need to receive an invite.

For this illustration, we're going to create a public team from scratch. We need to give our team a name. Again, you can add a description and then click on "Create."

It's going to create the team and now it's going to ask us who we want to add to this team. Once again you can start typing in the name of somebody from your organization then click on the "Add" button to add those members. Now at this stage, you can choose to make any of these people an owner and bump up their access a little or leave them all on members. We'll click on "Close."

Now we can see at the bottom we have our new team, and again, we have a general channel so we can start having conversations.

From template

The final way to create a team is to use a template. For this illustration we're going to click this template that says "Onboard Employees" and we see the channels that it's going to give us as part of this template - we're going to have General, Announcements, Employee Chat, and Training. It's also going to add nine applications for us.

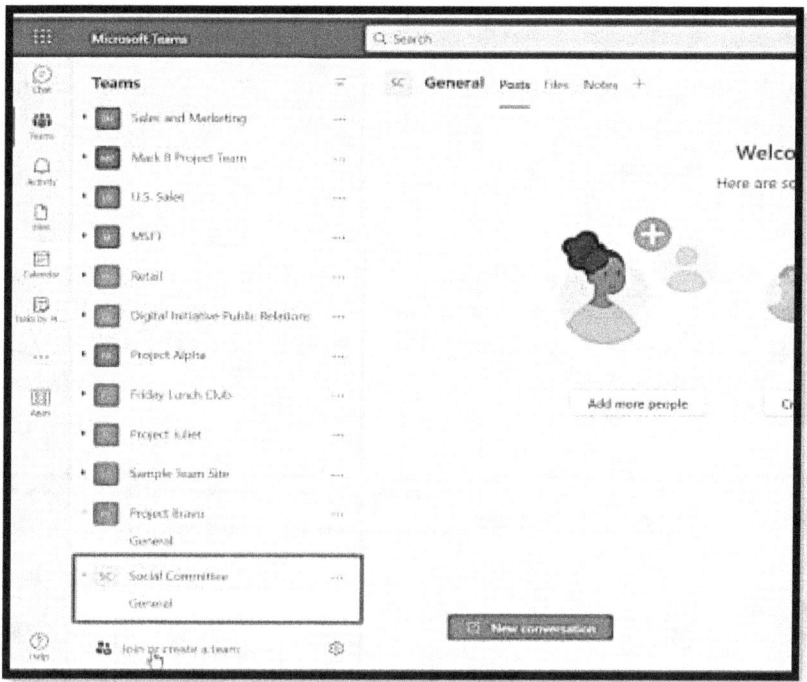

We'll click on next and for this illustration we're going to make this one a public team as well and now we can give our team a name. If you expand customized channels you can go in and rename these channels so you're not stuck with those defaults. After that, we'll click on "Create" and it's going to create the team. Remember, we're creating a public team again.

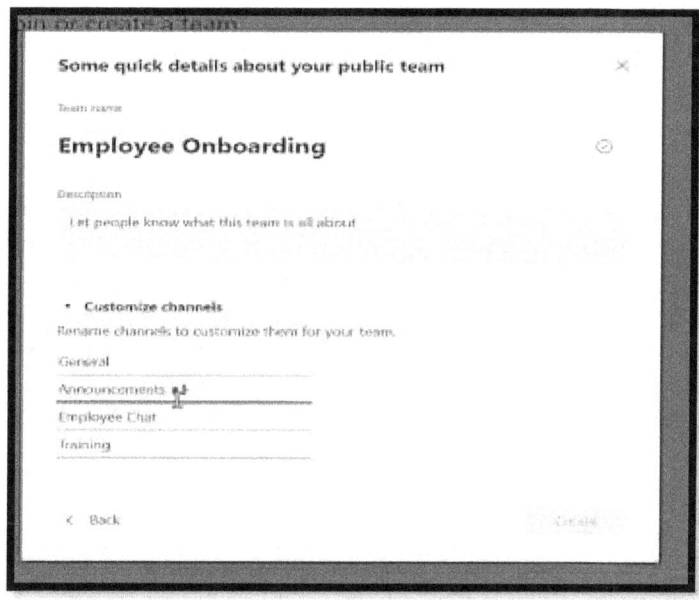

When you're using a template like this, it will take a bit longer to create the team simply because it's adding in lots of things like connectors to other applications and it's creating all of the channels so you can close this window and just carry on working and you'll be informed when that team has been created. You can see at the bottom here that Employee Onboarding has been created.

Joining a team

Going back into the "Join or create a team" we have something different here. Let's say we now have a team called Music and Movie recommendations and this is a public team that's been created by somebody else within our organization so this is what you'll see when somebody creates a public team - you will see it appear in this window just here. If you decide that you want to be part of this team all you need to do is hover over and click "Join team."

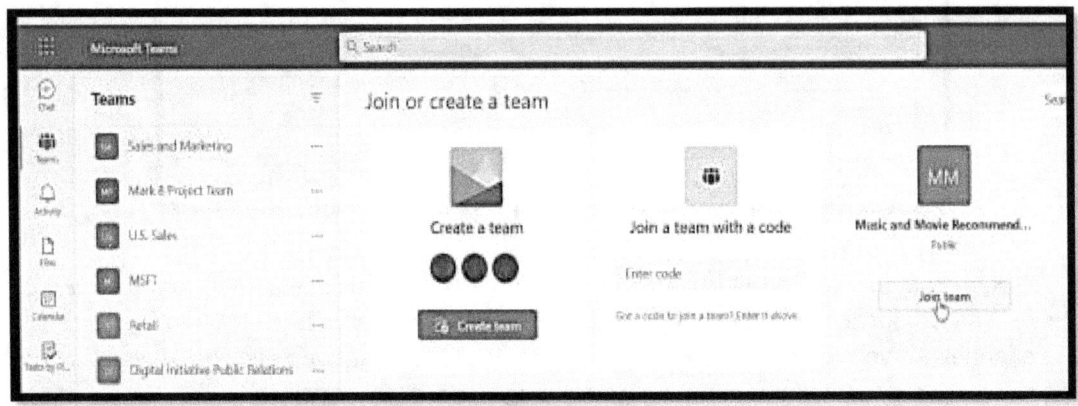

Now we've joined that team, if we take a look at our team's list we can see there is the team at the bottom for Music and Movie recommendations. That's how you can join a public team but joining a private team is slightly different from a private team. You need to be invited to join by the team owner - you'll get a pop-up from someone who's inviting you to the team. Now aside from the pop-up in the corner, you'll notice in the activity area you now have an alert or a notification and if you click on this it'll tell you that someone has added you to the team so now when you go to your Teams and scroll to the bottom you'll see you have access to that team. The other way you can join a team is by simply using a code. A code is something that can be given to you by the team owner. Let's show you an example of this. If we go to one of our teams, click the three dots, and go into Manage Team, underneath the Settings tab we have a team code area and what we can do here is generate a unique Team Code which we can then share with others to allow them to join the team. We could copy this code, paste it into a Channel or an email, and then those people can simply go to the "Join or Create team" option and type the code here to join the team.

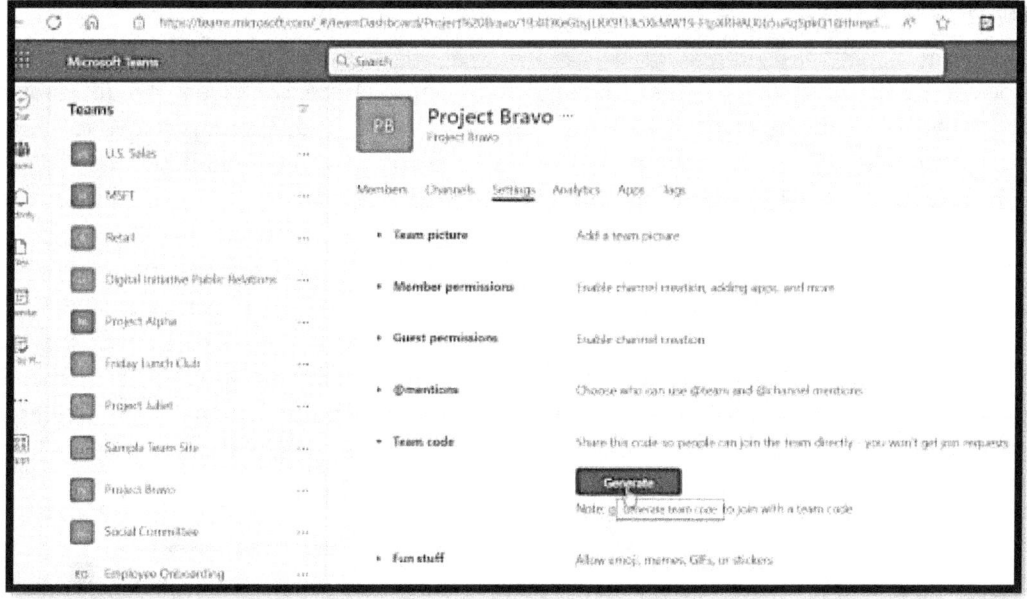

You can see there are a few different methods when it comes to creating teams and also joining public and private teams.

Managing Teams and Channels

It's important to know some of the options that you have when it comes to managing teams and channels within Microsoft Teams. For example, you're currently in the general channel in a team, at this point, you should be getting used to the fact that when you see three dots that means you have a menu that's going to show us some more options. If you click the three dots here you can see you have different items that will help you manage this team.

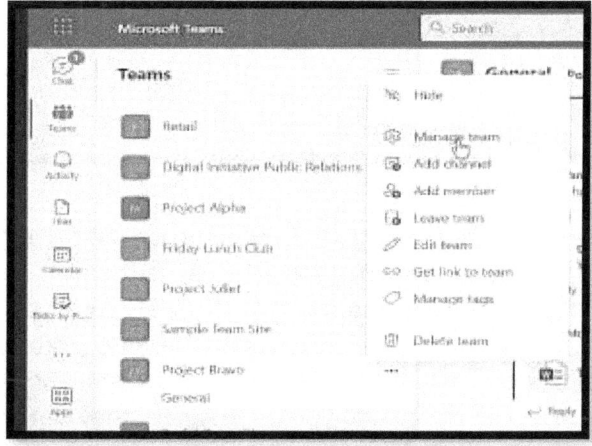

Managing teams

Let's look at the first one, which is the "Manage team" option. Here, we have a few different tabs running across the top. The first tab we have is the Members tab so if you ever need to see who exactly is a member of this particular team this is where you can come and it helpfully divides it down into owners, members, and guests. From here, you can see the owners and members. You've already seen how you can promote or demote these people so if as an owner you want to demote your co-owner back down to member status, you can simply do that in the column. You also have a search bar at the top. If you're in a team that has a lot of members you can search for specific members and if you hover your mouse over any of the members of this team you get a little pop-up window with a preview with different actions you can take. From here you can start a chat with a member, you can start a video call, you can also send them a quick message from here as well and you have access to all of their contact details. This can be handy. You can also add new members using the add member link in the top corner.

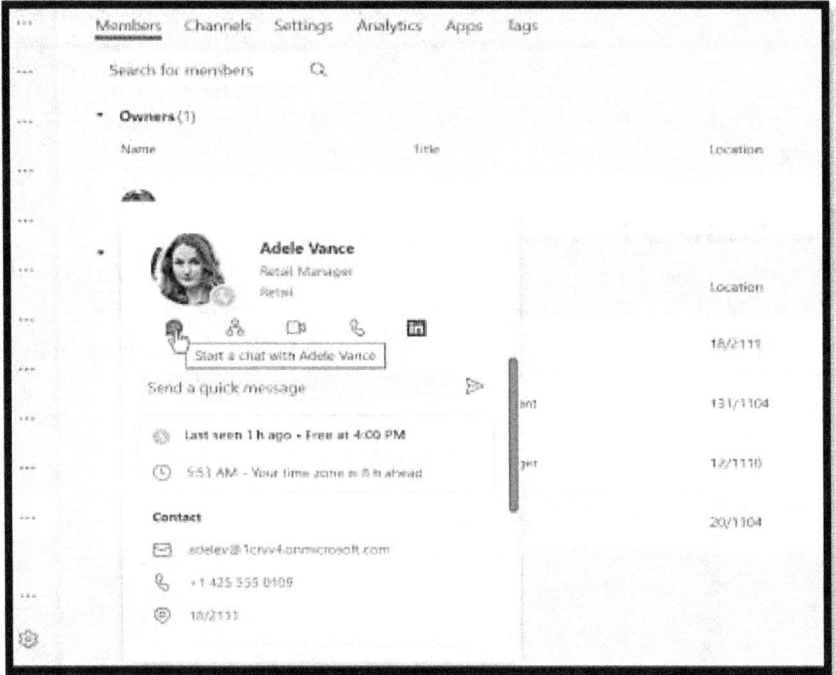

Let's go back into the Manage team and go to the Settings page. This is where you can customize all of the settings for your team. For example, you can change this avatar from just the standard fairly boring default Avatar to a different picture. Click "Change picture" upload a picture and click on Save. You can also set member permissions. This is a series of checkboxes that you can toggle off and on that enable or disable certain permissions for this team. For example, we have all of these toggled on so members can create and update channels, they can create private channels, delete and restore channels, and so on. Have a little look through these because it might be that you don't want members to be able to do certain things such as deleting and restoring channels and if that is the case, then you could deselect that option so be sure to have a check and modify when necessary. You also have a section for guest permission and this is all about what you want to allow guests to do.

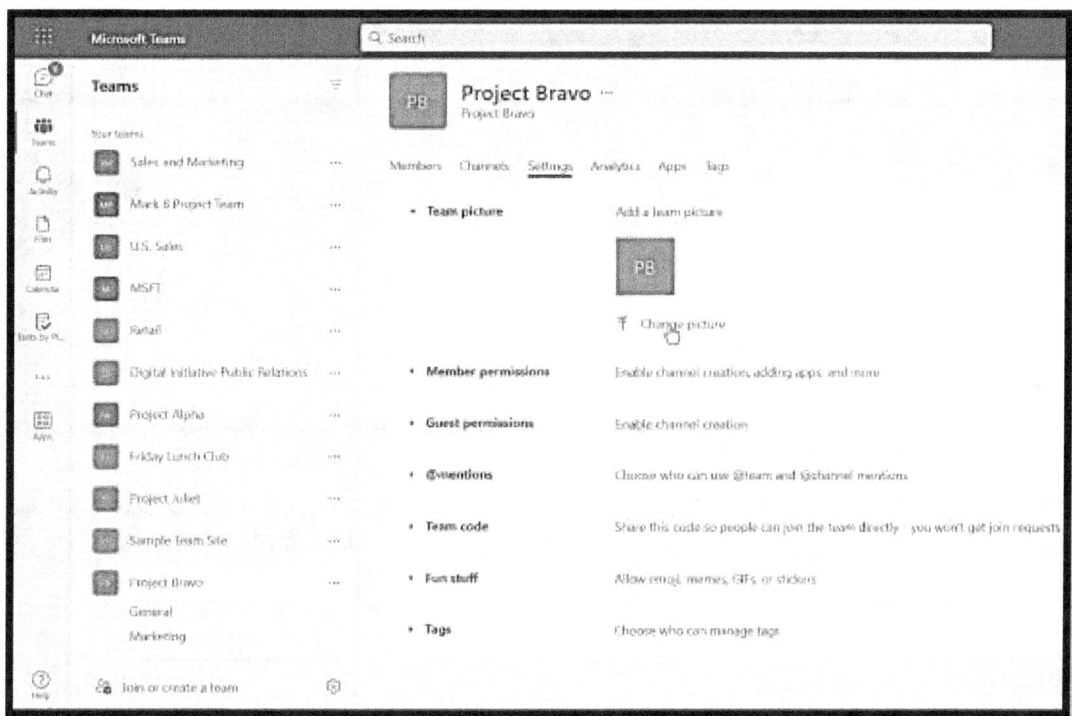

You have an @mention section and this is an area where you can define who can use @team and @channel mentions because remember you can @mention other people but you can also @mention an entire team to send a message to all of those team members and you can also do same for a specific Channel as well. For the Team code area we looked at earlier but this is where you have a unique code related to this team that you can send to other people to join the team. You have a "Copy" button here which is an easy way to copy it into an email. The Fun Stuff area is where you have things related to emojis, memes, gifs, and stickers so have a little look through there. By default, everything here is selected. The final section at the bottom is related to tags and in this section; you can choose who can manage all the tags related to this team. You can see that currently tags are managed by the team owners. If you carry on moving across these tabs, you have an Analytics tab which comes in handy if you want to see an overview or some statistics about your team. You have an Apps tab that will show you all of the apps that you currently have installed.

Creating Tags

You have a Tags section and what you can do here is create a tag. Tags allow you to quickly reach a group of people all at once. For example, if you have numerous people that have the same job title, let's say sales assistant, you could set up a tag for sales assistant and it will effectively group all of those.

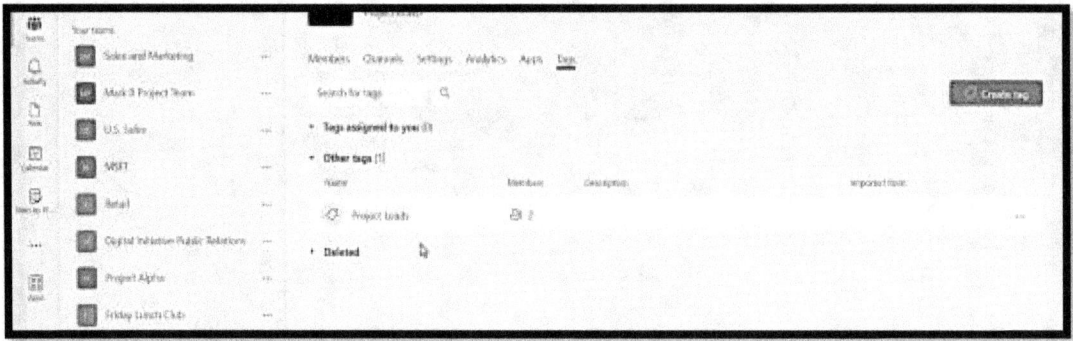

Let's create a tag called Project Leads, add a description, and then we can add people to this tag. Once we've created these tags we can then use them very similar to an @mention in posts and this Tags tab is where you can come to manage your tags and create new tags. Those are all the team settings but it's worth noting that we also have channel settings.

Channels

Inside a channel, you've got a little page that has posts. They're like a little message board that you can pop up with different things that the whole team needs to know. You're going to click to start a post and here you can add a subject to the post and then type the message underneath. There are some settings here where you can choose to post this in multiple channels or just this channel then you can choose that everyone can reply or just you and the moderators can reply. There are a few more options that you have inside of here but just know that chats are with you and one other person; it's more like direct communication. This is more of team announcements and team communication so if somebody has sent you an invite to a team and you haven't seen it yet you can click on those three dots above and click Pending invites. When people invite your team to a shared Channel, the invite will show up there then you can also choose on those three dots the Analytics section. You can see how many people are active inside of your team, how many posts and how many replies there are, and the type of channel, whether it's private or public. Inside Teams you have a lot of options inside the settings of the team. When you click on "Manage team" and you go to the Settings tab, there are a lot of options to remove permissions for members, remove or add things to the channel like memes, and stickers, or also allow members to add or remove apps, or upload custom apps. You can also allow members to create private channels or not.

You have the option to create private channels under a team. This means you need to have an invite to be able to get to them because just like there's a private and a public team, you can have private and public channels underneath that team. Something to note is that Microsoft Teams does connect well with SharePoint so every time you create a team under Microsoft Teams, it does create a SharePoint site. We suggest you connect them because it's so efficient for your team and keep files in places where only the people you want can access them and the people that you don't cannot access them.

Now if you jump across to the Channels tab at the top this is where you're going to see all of the channels that are part of this team. You'll get to see the general channel and then you can add new channels from here as well. If you click on the "Add Channel" button you can create a brand new channel for this team. Let's say you need a marketing channel, you could add a description and then keep this on standard access so everyone on the team has access to this channel. Clicking the drop-down brings up a few different options that you can pick in there and after all that is done, you will see that this channel has been added. If you click the three dots next to a channel you have other options here as well. For example, you can manage your channels, you can set the permissions, you can set up moderation on channels if you want to and you also have access to some analytics for the channel.

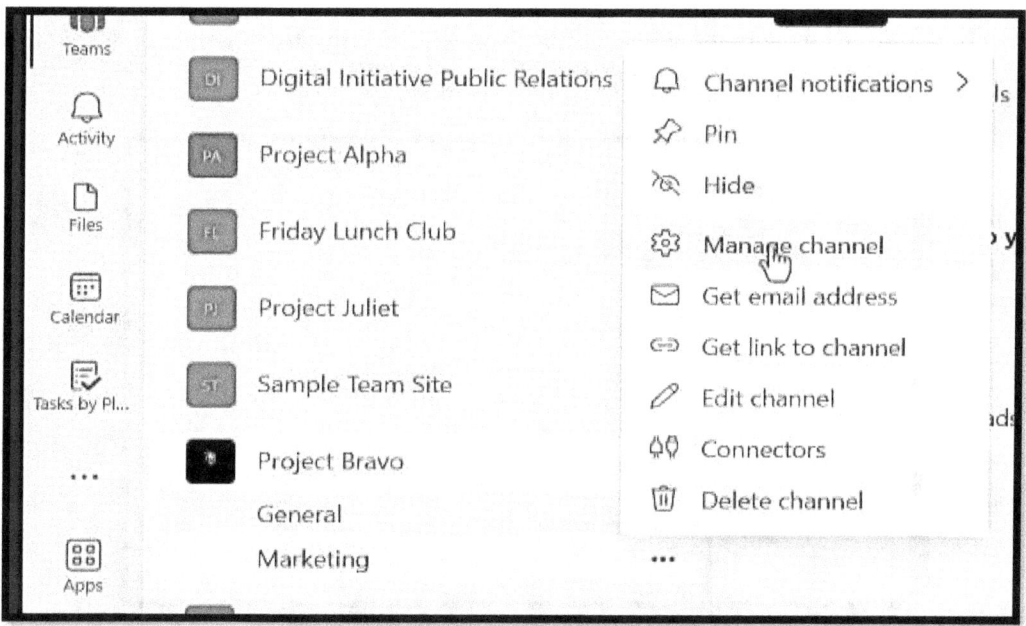

The final point worth mentioning is that you don't necessarily have to go into the "Manage team" area to create things like New channels; you can simply do that by clicking the three dots where you also have an "Add Channel" option here.

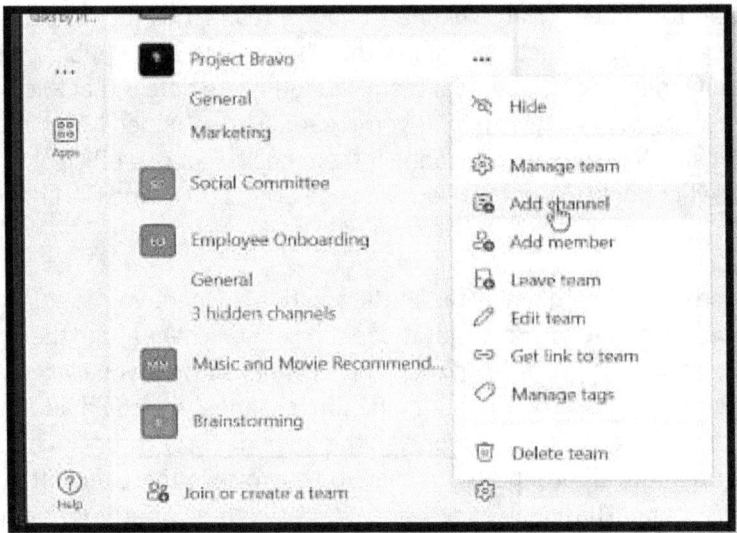

Sometimes when we create teams we create them because they're related to a specific project and at some point in the future that project will end and we no longer need to be part of this team so in Teams you have three options: you can leave a team, hide a team or delete a team. Note that the final option to delete a team is only available to whoever the team owner is, which means members of a team can't delete the team.

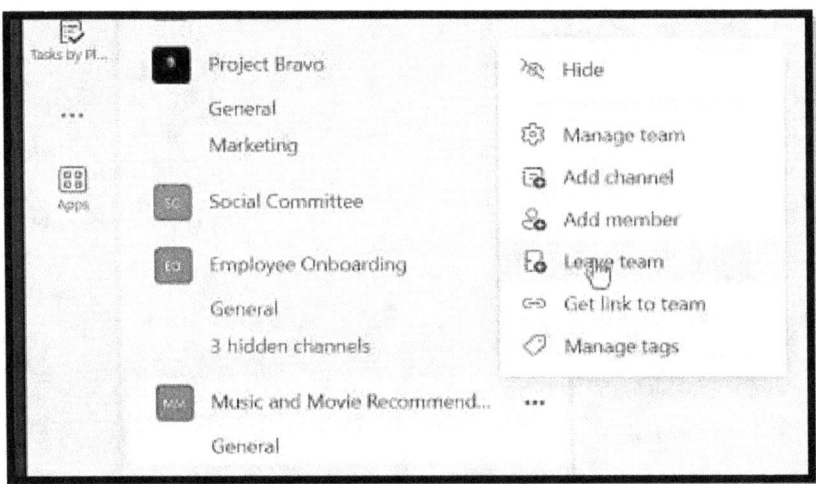

Leaving a team

Let's talk about leaving a team. For this, we'll look at a team that you didn't create. Let's say you're not interested in having this team on your list and you want to leave the team. This is a simple case of clicking the three dots, clicking the "Leave team" option, confirming that you want to leave and you'll notice that that team gets removed from your team's list. It is pretty much as simple as that.

Rejoining a team you left

Now if you decide that you want to rejoin that team and you go back to the "Join or Create a team" option, because this is a public team you can see it there. Just click "Join team" and that will put it back into your team's list.

Hiding a team

The next thing you can do is hide a team. Let's say you don't want to leave or delete the team but you just want to hide it from your list because you've got quite a few teams here and you only want to show the ones that you use every day. Once again, go to the team, click on the three dots, click on the "Hide" option and it's going to hide that team so it disappears from the list. You can see right at the bottom you have a little hidden team section and expanding this is going to show you any of the teams that you've hidden and if you want to unhide them simply click on the three dots and just choose "Show" to pull that back into the main teams list.

Deleting a team

The final option that you have is to delete a team. As we mentioned you can only delete teams that you've created. If you go to a team that wasn't created by you and you click on the three dots, you'll notice that you don't have the option to Delete the team but if you did create that team and you want to delete it, click on the three dots at the top and you have a "Delete team" option at the bottom.

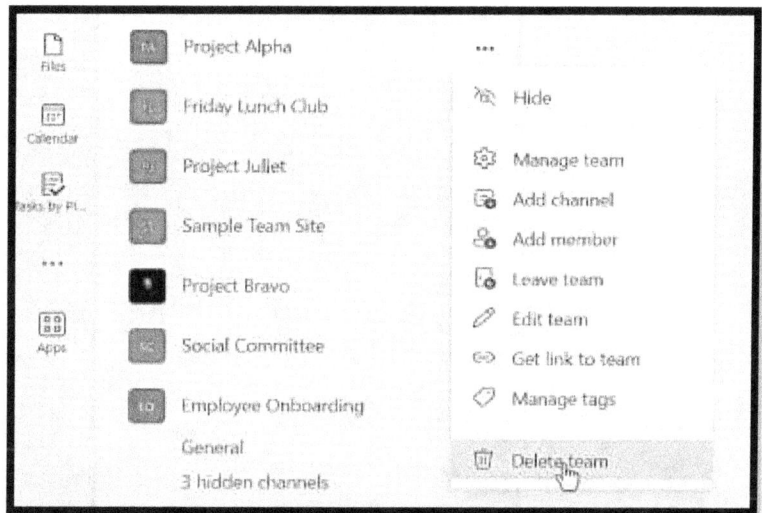

Click on that option, check the checkbox to confirm that you understand that everything is going to be deleted, and click on "Delete team." This, together with all of its contents will be removed. Note that it doesn't just delete the team for you, it will delete the team for everybody who has access to that team so just be aware of that; sometimes it's preferable to hide the team if you just want to remove it from your list as opposed to deleting it. Now that you have a good understanding of how many different elements work in Teams it's time to turn our attention towards adding tabs at the top and also searching.

Review Questions

1. Create a new team using a predefined template.
2. Create a new channel within an existing team.
3. Observe how team members can be added removed, or have their roles changed within the team.

CHAPTER 4
TEAMS CHATS AND CONVERSATIONS

Chats are conversations with you and one other person or you and two or three other people but we would recommend you don't make it much bigger than that; it's supposed to be just direct communication between you and somebody else and should be quick and easy to do like a text message.

Teams Chats

If you want to create a new chat go to the top right menu and click "New chat."

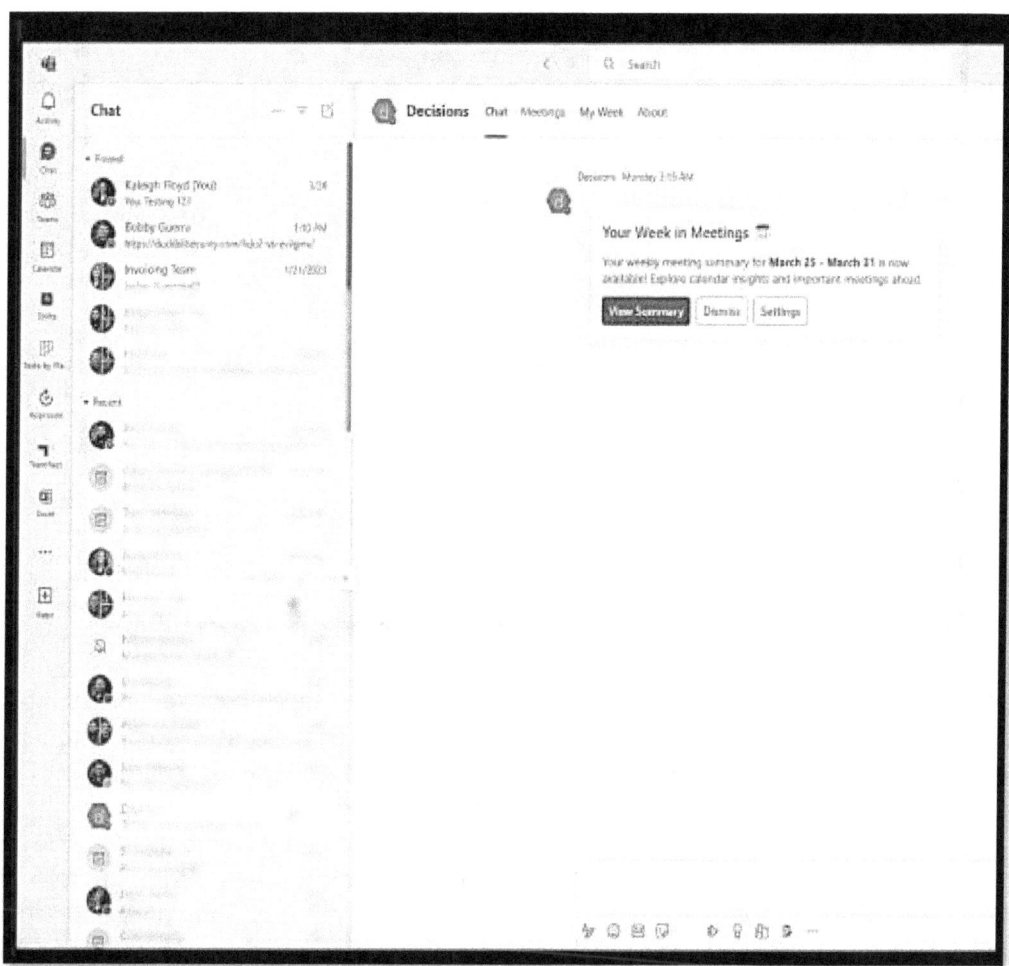

This creates a new entire window of itself and you'll notice how it broke out this cute little window. Here you can choose who you want to talk to and then you can go ahead and start typing to them. If you click on these three dots next to where you just created a new chat, you have more options. You can choose to mark all as read which is something that works if you want to get rid of all of your notifications. You'll also notice you have some chats at the top and then others at the bottom. The ones up top are your pinned chats and you can pin a chat by right-clicking on that chat and clicking "Pin." When you right-click you'll see you have a few other options as well. You can mute that chat, hide it, or manage apps inside of that chat, and at the very top you can choose to open that chat in a whole new window so when you click on that, you'll notice that your conversation is now in its own window in itself.

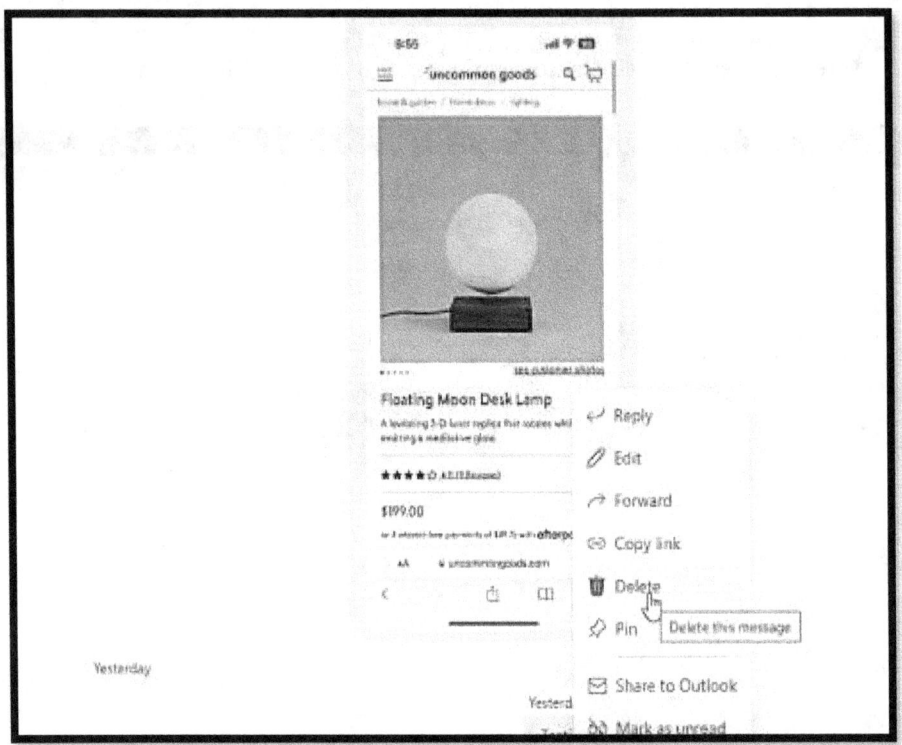

There's something you should note about chats and typing a message. If you're typing something in a Word document and you're going to the next line down, if you click Enter it should go down to the next line but when you click Enter inside of Microsoft Teams it just sends the message. There's a solution for that and that's by using the formatting button. So let's say you were typing something and wanted to go to the next line, you would click this format button right here (it's the little A with the pencil) and it makes the Box bigger then it gives tons of options when it comes to formatting this. When you click enter it goes down to the next line so if you're typing something and you want multiple lines we would highly suggest clicking that button. Another cool thing is you can add in things like cells and you can also Mark your message as important if you want to and if you want to discard it you just click the delete button at the top, say discard and it would be gone. Another thing to note is that in your message if you right-click on that you can choose to delete the message, and when you click on that it shows the person that the message has been deleted but it doesn't show the information to the other person. Also, if you right-click you can choose to pin that specific message, and you can copy the link to that message which copies an exact link to the chat and everything, you can also forward the message so it takes that message and you can choose to send it to somebody else with the exact information and who wrote that message to that other person. This could be helpful if you were just trying to copy some information to somebody and didn't feel like texting it out completely as you would just click forward, send it to whoever you want to send it to then and it would forward that message to them. Notice how it says the name of who wrote that message and then the message information below it. It also says the time the message was sent and the date that it was sent.

Posting and receiving messages

Posting and receiving messages in teams is a super simple task. As we already know we can see all of the teams that we have access to in this pane on the left-hand side and then when we expand a team you can see that we have all of the team channels now underneath that team. Remember, there will always be a general Channel created by default when you create a new team and this is one of the channels that we can have conversations in. We can create our own channels and we are going to take a look at that a bit later on but for the time being, let's just check out the options that we have when it comes to posting and receiving messages.

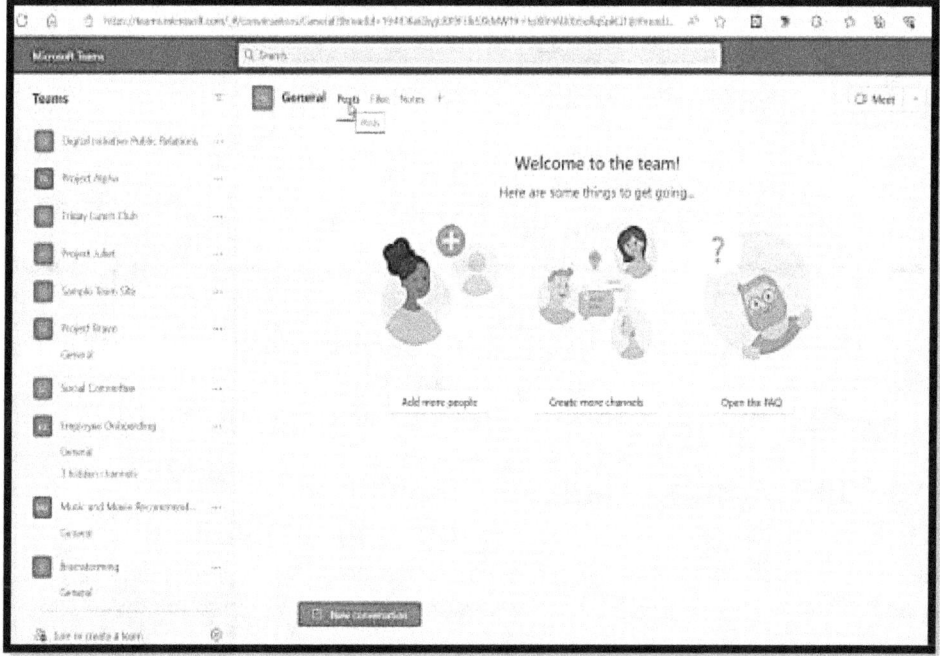

We are in the General Channel and the first thing you'll notice is that right at the top where we have the channel name, we also have a Posts tab this is where you can chat with your colleagues, send messages, attach files, post images, gifs, stickers, emojis and all of those kinds of fun things.

Starting a new conversation

To start a new conversation goes to the bottom of the screen and click the "New conversation" button. This is going to pop open a little conversation field with some icons underneath. You can type your message and now to send a message through you have two options: you can either hit the Enter key on the keyboard or you can click on the little paper plane icon on the right-hand side and as soon as you click that it's going to post the message to the channel and then your colleagues can come in to reply or they can post their own messages.

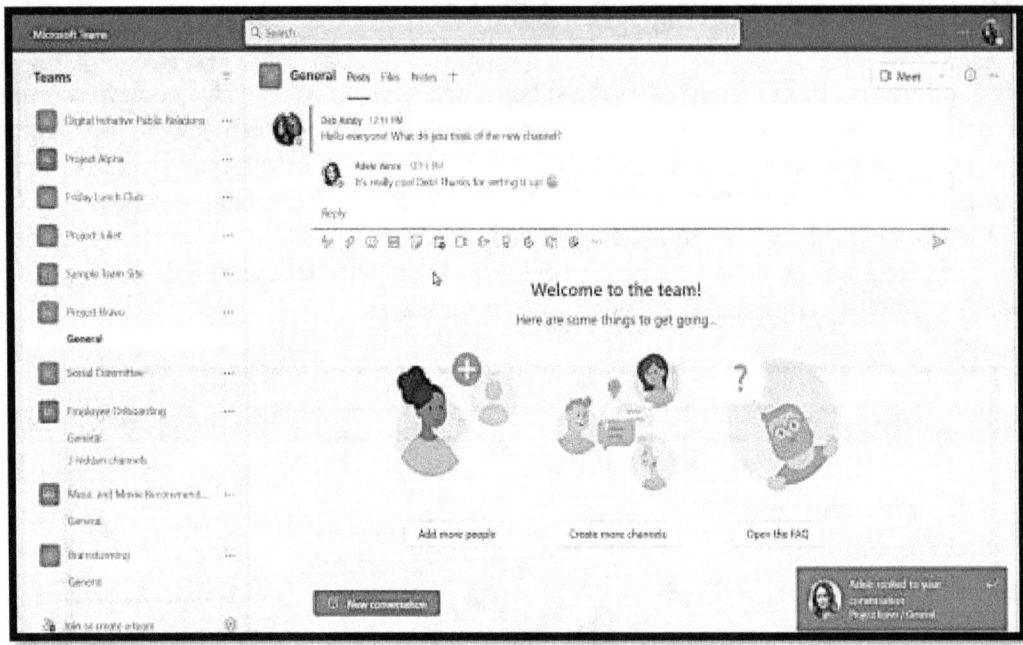

When you do receive a message you'll get a little pop-up notification in the bottom right-hand corner. Also, notice that when you're working with conversations in Teams you have some kind of threaded conversation layout so you can see very easily who's replying to what.

Adding fun to your chats

Now if you want to reply to someone, ensure you've clicked in the correct field then just type in your message. If you want to add a little bit of personality to this message you have some options underneath.

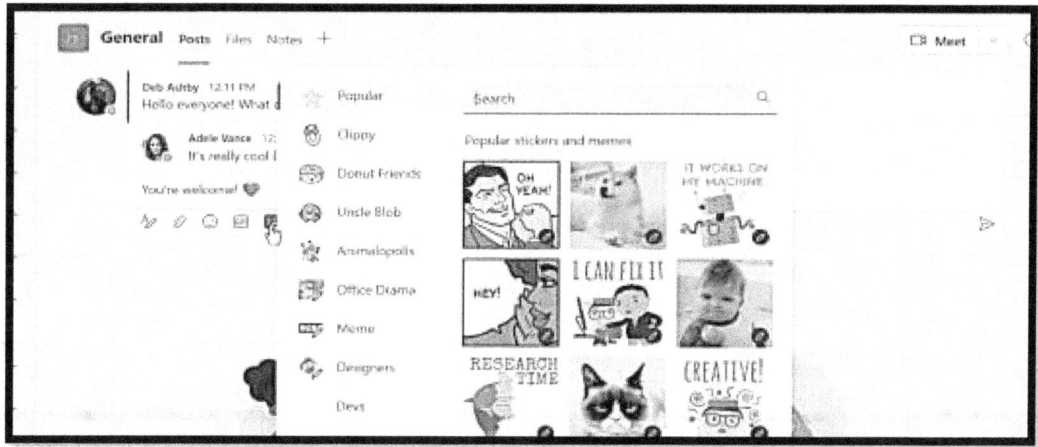

For example, you could add an emoji so if you click the emoji button you get a whole list of different smileys and other things that you can add. You can also add a GIF so if you click to open this up you have a whole host of little GIFs that you can add just to make your conversations a little bit more fun. The final option you have here for giving your messages a bit of personality is the sticker's option so again, these are divided into different categories and these are just fun little things that you can send through. After adding whichever one you want, click on the Send button and that's going to be sent through to the other person and that person is going to get a pop-up notification in their teams that you've responded to their message. There are times when you'll see here that the person has just sent through another brand new message and hasn't responded to the previous conversation thread; instead, they started a whole new post. You'll also notice when you hover over the person's message you get a little toolbar that pops up where you can add a reaction to their post so if you like this post you can simply give it a little like by clicking on the thumbs up icon.

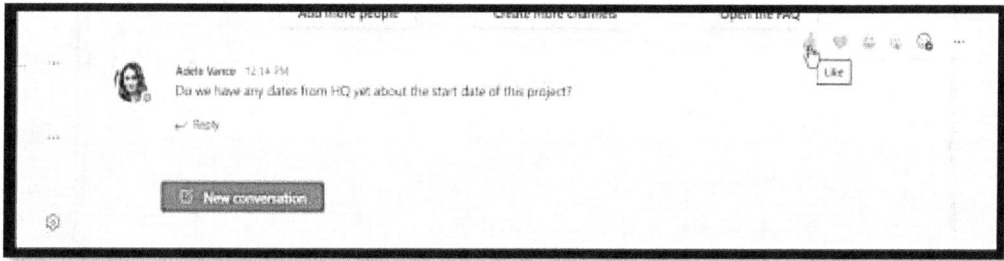

Formatting your chats

Now you can reply to their messages again but this time you're going to click on the format message button underneath. What you'll notice is that it opens up a bigger reply window and in this window; you have all of your formatting options.

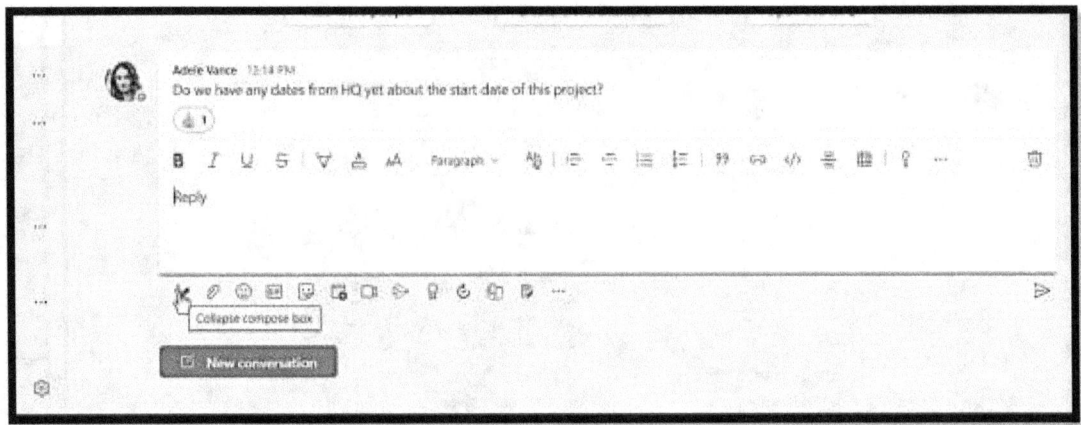

If you want to add a bit of formatting to your message as opposed to just having it all in the same font and the same style then just click on that format button. You can make your letters bold, change the color of the font, and even do things like add bullet points, quotes, tables, and at any point when composing a message if you decide that you no longer want to send this you have a trash can icon on the right-hand side which will allow you to effectively delete this draft.

Sharing and accessing files via chat

Also, don't forget that you can share files here as well. If you start another new conversation, notice you have a paperclip icon just here so if you click this you get a few different options: you can browse your recent files, browse other teams and channels for files that have been shared there, you can go directly to OneDrive or upload a file from your computer.

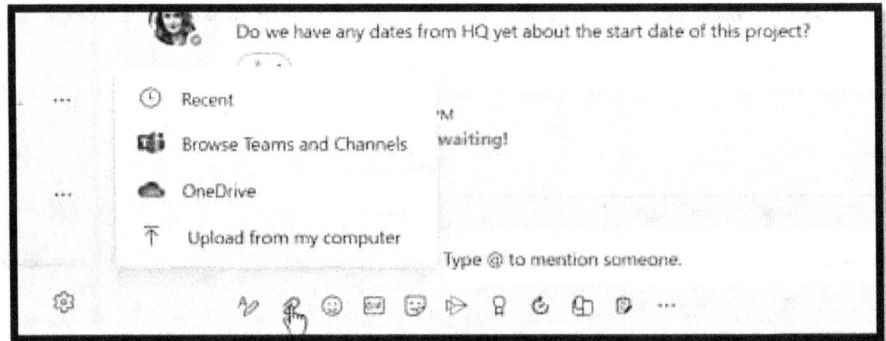

If you go to OneDrive, you can go to a Files folder and send documents from there. As you click on that you'll get two options: you can upload a copy or share a link. Let's say you choose to upload a copy, it's going to load that into the message and then you can send it through as you normally would so anybody in this channel can then open this document.

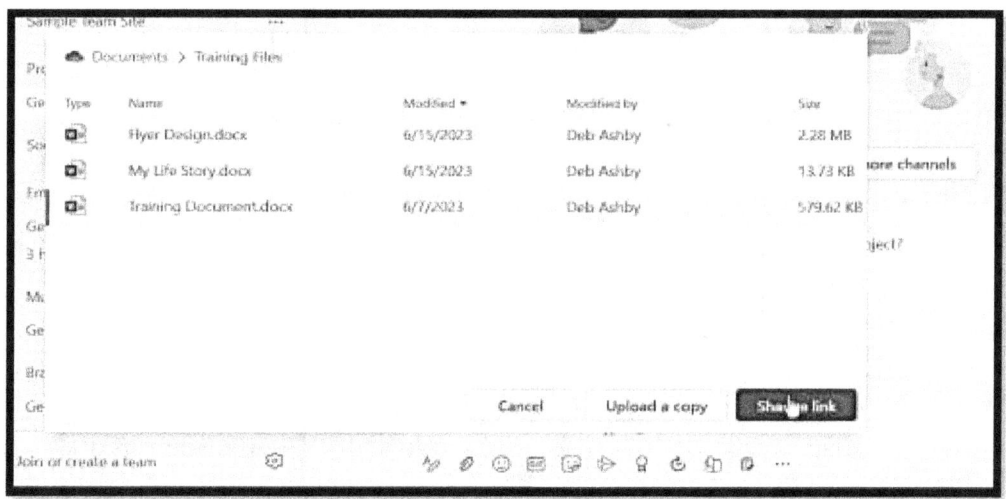

It's also worth noting that when you share documents in this post area they upload and become available in the Files tab at the top so if you quickly need to see all of the files that you shared within this channel you can just go to Files and you'll find all of them sitting there. It just means you don't have to scroll up and down a conversation looking for your files.

Addressing a specific person

You can also use @ mentions in your conversations. Let's say you want to reply to a post again and you want to bring someone into the conversation. You can just type @ and then you can either select from the list or start to type their names and choose the person. After that, type in your message, hit enters to send that through and that's pretty much all there is to sending and receiving messages.

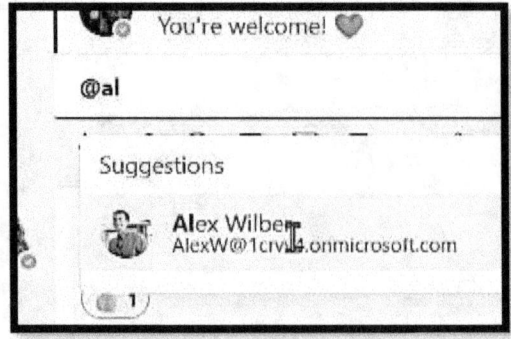

Remember with these @ mentions you don't necessarily need to have the full name just there, sometimes that sounds a little bit formal so once you've added them in you can backspace to remove that last name so it just says Alex, instead of Alex Morgan for instance. If you've used any applications like this previously you're not going to find this too hard to grasp.

Bookmarking, editing, and deleting posts

In this section, we're going to take a look at bookmarking, editing, and deleting posts. Bookmarking is a great way to be able to essentially flag messages that you might want to come back to later on. It's similar to bookmarking a website that you're interested in reading. The way that we bookmark posts in Teams is not immediately obvious. Let's look at a channel that has a little bit more in it. Let's say we want to bookmark a message from someone, what we can do is hover the mouse over this message, and where we have all of our reaction icons, you'll notice that next to that we have three dots for more options. If we click more options we have an item in this menu called "Save this message" so it's not called bookmark but bookmark accurately describes what you're doing.

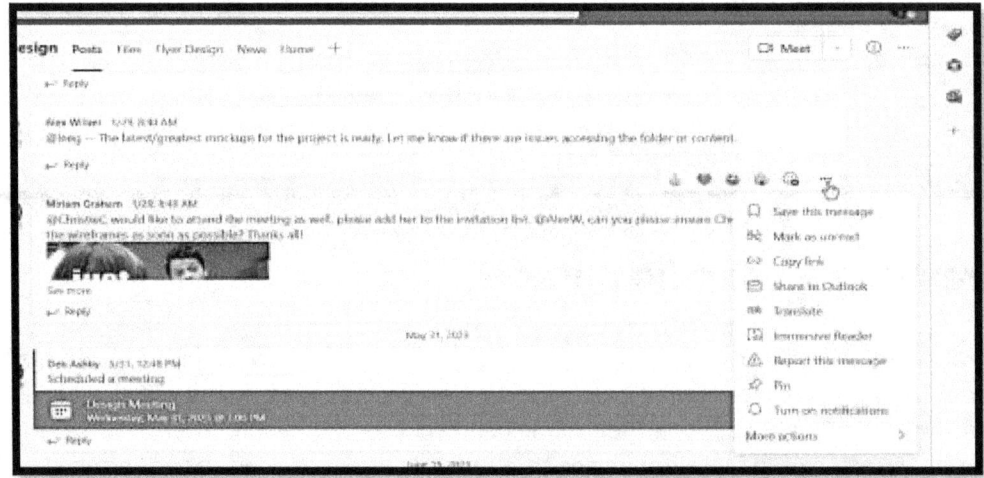

If we click on "Save this message" we get the notification that it's been saved and then when we want to access all of our saved messages we need to go to where we have our profile picture, click on it and there we have our "Saved" option.

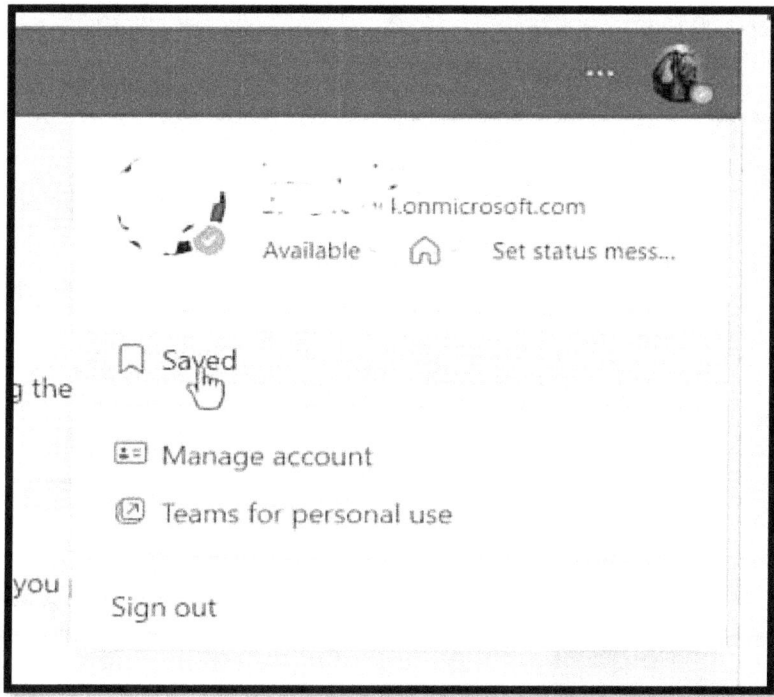

If we click on "Saved" this is going to show us a long list of all of those messages that we've bookmarked and of course, if you want to unsave a message you can just click on the little Saved icon and that will remove it from your saved items list.

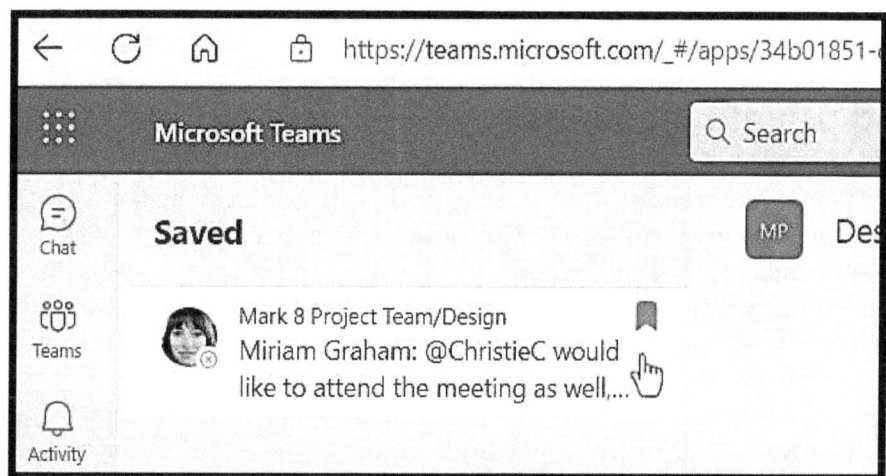

Now let's talk about how we can edit messages. Let's say we have a message we want to edit, again, this is a simple case of hovering over the message that you want to edit and it is worth noting you can only edit your own messages. Once again, we'll go back to the three dots and we have an "Edit" option in this menu. If we click on that it puts us in Edit mode and we can make changes, and then click on the tick to send that through, and of course, the receiver is going to get a notification that this message has been updated so they can go back in and check it.

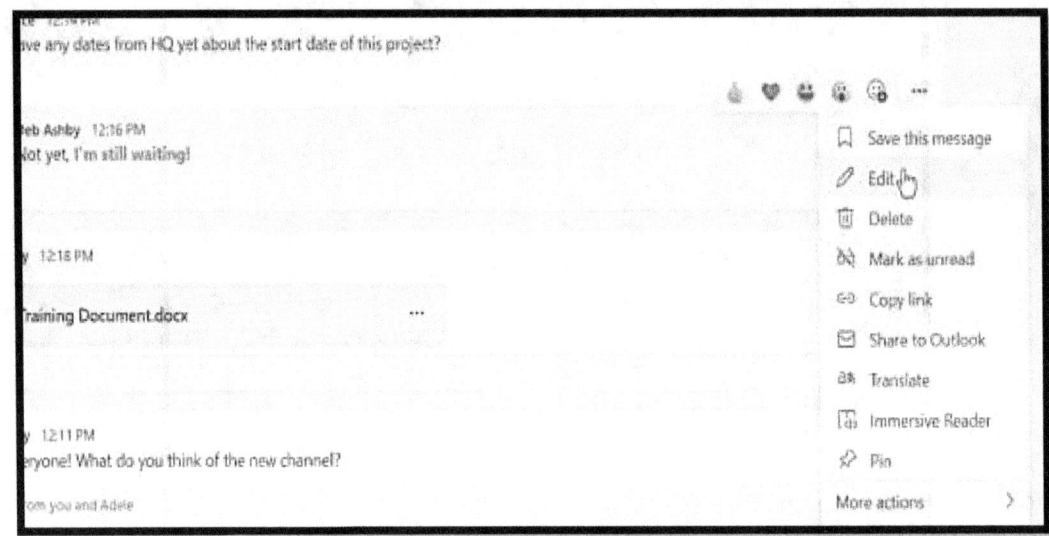

Lastly, let's talk about how we can delete messages. As you would imagine, this is pretty standard and straightforward too. Let's say that we want to delete a message here. Once again, we'll hover over the message, go to the three dots, and in there we have a Delete option. As we click on Delete, we'll get a little message that lets us know that this message has been deleted. Note that you're the only one that can see this message so everybody else in the channel is not going to know that you've deleted this post. If you decide that you want to undo that and put the message back, you have an undo button just there which will pull that message back into the conversation.

Sending private messages

We have been looking at messages that we share with all of our team members and the people that can see all of these messages are all of the people that have access to this Channel or all the people that are members of this channel. Sometimes you might not want to post a message to the entire group, you might just want to post a message or have a conversation with a specific team colleague and this is what we refer to as private messaging. Let's say a colleague was off work yesterday and you just want to have a quick conversation with her to make sure that she's fine, this isn't a conversation that's appropriate for everybody on the team. It's just something that needs to occur privately between us so for private messages you don't type those into the team channel. To do this go to the Chat area of Teams. If you take a look at the left-hand menu you can see the first option you have is chat and this is where all of your private messages occur. In this window, you can see a list of the recent conversations that you've had with people so you can carry on typing in those or you can create a new chat at the top.

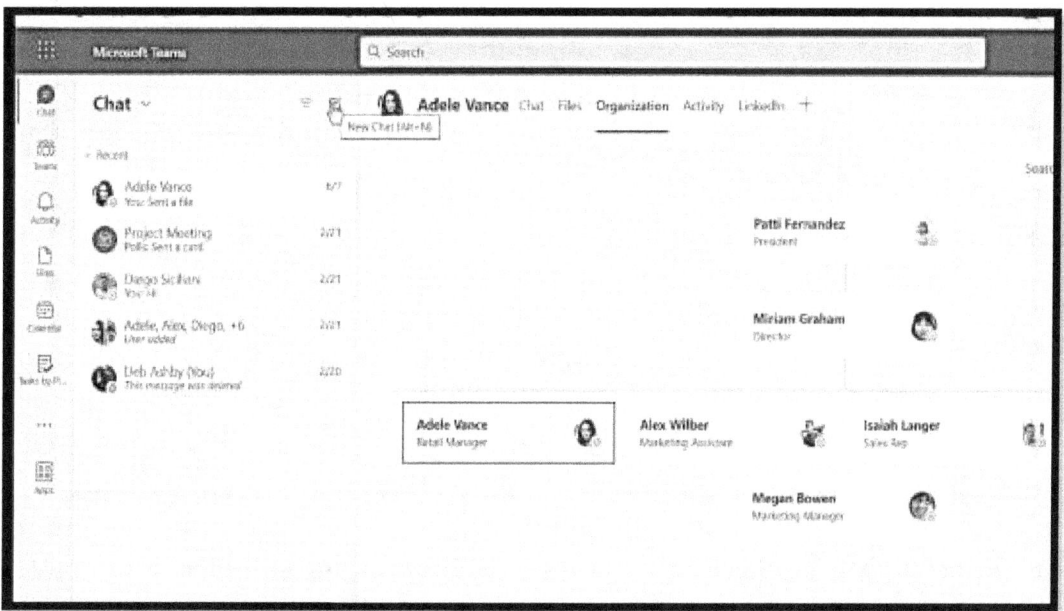

If you click on "New chat" you can enter who you're sending this private message to, so once again you can type in the name or you can select from the list if they're within your organization. You can have a private chat with more than one person; it's not just a one-to-one thing, you could add in Mary and John or whoever you want. Let's say you want to message Mary. Now if you've already had private messages with Mary before, as soon as it recognizes that you've put Mary in that "To" it's going to jump you to the last conversation you have so you're not clogging up your recent area with brand new chats with the same person. Now you can simply type in your message and then click on the send icon to send that through and Mary is the only one who's going to see and reply to that. One other thing to note when you're working within private chats is that you also have different tabs running across the top. You can customize some of these but you'll always have a chat Tab and you'll also have a Files tab. This works very similar to when you're having conversations in a team's channel. It's going to show any files that have been shared within this private chat so just be aware of that.

You can add other tabs across the top which is why you can see "More" in this private conversation. It's also worth noting if at this stage you decide that you want to jump into a video call with this person and have a conversation with them, in the top right-hand corner you have icons that allow you to do exactly that. You can choose to start a video call with this person or start an audio call. We're going to be taking a look at video and audio calls later but that's pretty much how private chat works.

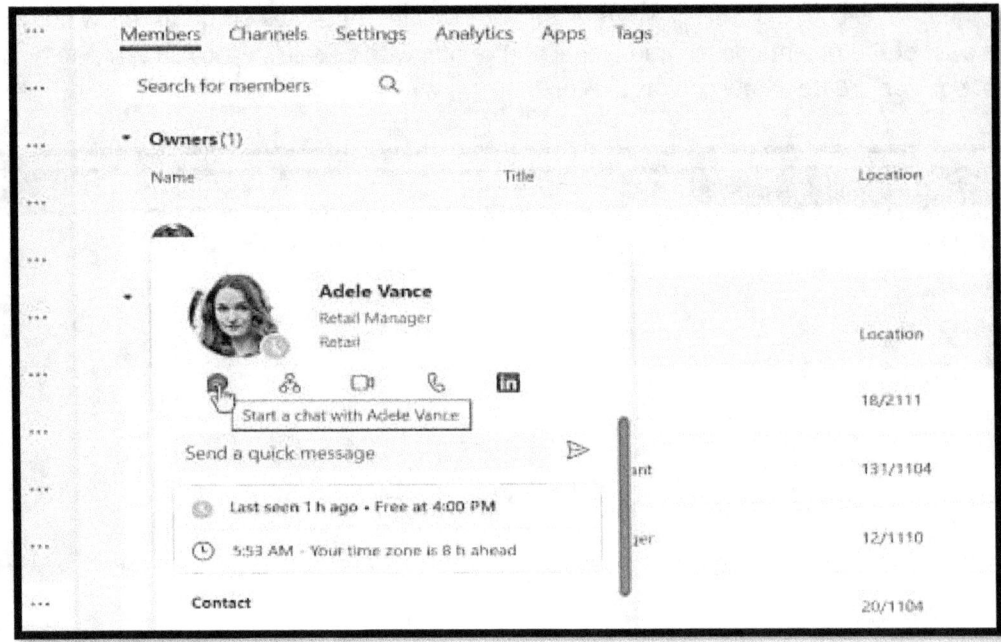

The only other thing to mention here is that you do have a little filter at the top so if you click on the filter you can filter by name. Let's say you're looking for conversations that you've had with Diego, you can type in his name and it's going to show you all of the conversations that he's been a part of so don't forget about that at the top. Now the very final point to mention here is that if you hover your mouse over any private conversation, again you have those three dots that take **you to more options so you have a few different things you can do here:**

- You could pin this conversation to the top. This is useful if you have private conversations with the same person fairly often to make them easier to access.
- You can mark this conversation as unread. As soon as you do that it puts this conversation into a bold font as if you haven't read it yet and you can also see that on the chat icon in that left-hand menu, you now get a little notification letting you know that you have an unread conversation.
- You can also do things like completely mute the chat. Let's say you're trying to do some focused work, you can mute the conversation.

A good option here is to notify when available. If you take a look at your profile photo at the top you'll notice you have a green tick. That means that you're available and you can see that underneath your username but then you can change that at any point in time and this means changing your availability status. Sometimes these will be changed automatically depending on what activity you have going on in your calendar. For example, if Teams recognize that you've got a meeting scheduled it will automatically set you to busy but you can also manually change your status if you set your status to "Be right back" you can see that it changes the icon just there and if you know that you're going to be away for let's say, an hour, you could even specify the duration that you want this status to be applied. For example, you could Select 1 hour from the list here, click on "Done" and then your status is going to say "Be right back for the next hour" after which it will automatically revert to available. Now let's say someone has done this on theirs and you have something important that you need to talk to them about, this is where the "Notify when available" option can be useful. If you set this on this private conversation with this person, as soon as their status comes back to Available you're going to get a notification and you can then start your chat with them.

Review Questions

1. Initiate a new chat conversation with a colleague in the Teams app.
2. Practice formatting your chat messages using different styles and emojis.
3. Explore the options for sharing files and other content within a chat.

CHAPTER 5
SETUP MEETINGS AND CALLS

We've talked about teams, channels, and conversations. Now let's move on to talking about initiating video and audio calls because these days that is one of the primary uses of Microsoft Teams. Since the pandemic when everybody moved online, Teams is now used heavily for video and audio meetings, sharing files, and collaborating with other people.

Before you initiate video and audio calls

Before we look at how you can initiate video and audio calls, it's worth checking a few settings for your team meetings. If you look at the top right-hand corner where we have our profile picture, you'll notice we have three dots next to the profile. Clicking on this will give you access to your settings.

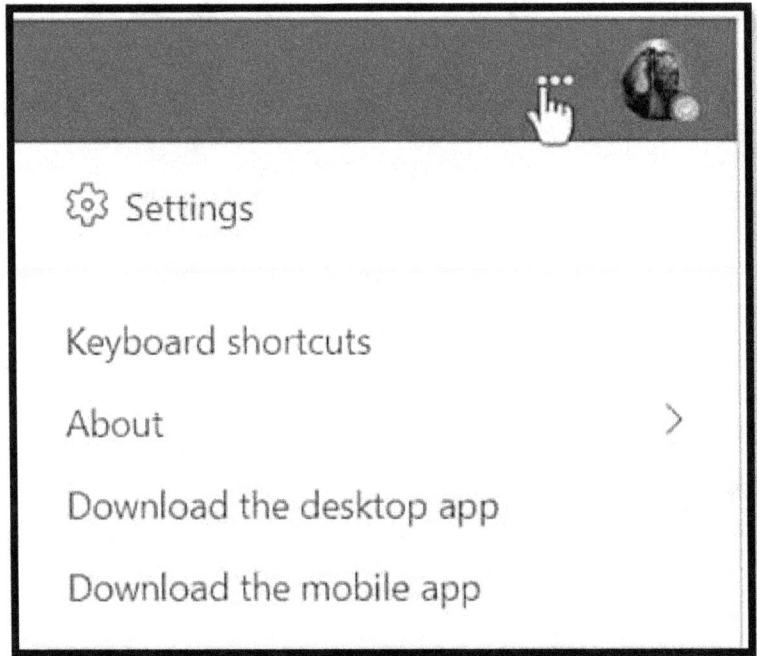

Now these are settings that apply to every aspect of teams and we're not going to go through all of these; we'll just go through the ones that are relevant to video and audio meetings but we would highly encourage you to have a little look through all of these settings and modify them and customize them the way that your teams is set up so it works best for you.

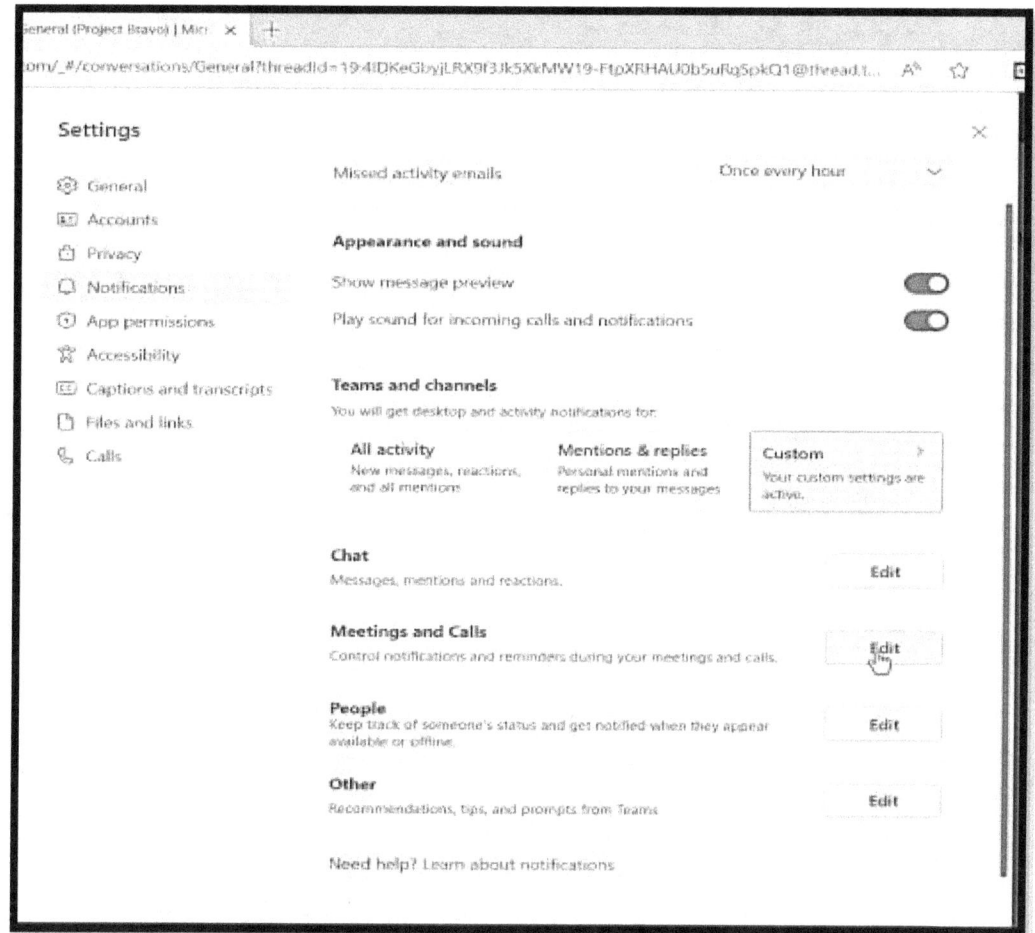

Let's go down to Notifications because we have a couple of things here that relate to Teams' meetings. If you scroll down to the section called Meetings and Calls this is where you can control notifications and reminders during and after meetings and calls. If you click on the Edit button you can customize your behavior when you join a meeting. For example, if you want all of the participants to have their microphones on mute you could choose this option. You can unmute them all so they can all have a chat or you can mute them until you join as the meeting organizer or send a message. This second option here is the default so just be aware of that when you're hosting webinars or Live events. Let's go down to Accessibility because we have another option here that relates specifically to Teams' meetings and that is Captions. You can see here you have the option always to show captions in your meetings. You can have this toggled off but if you toggle this on, this means that captions automatically pop up on the screen when anyone is talking in the meeting which is greatly going to help anybody who may have some hearing difficulties or for people who don't have English as their first language because sometimes people find it easier to read words than trying to understand different accents and voices so captions are always a good option.

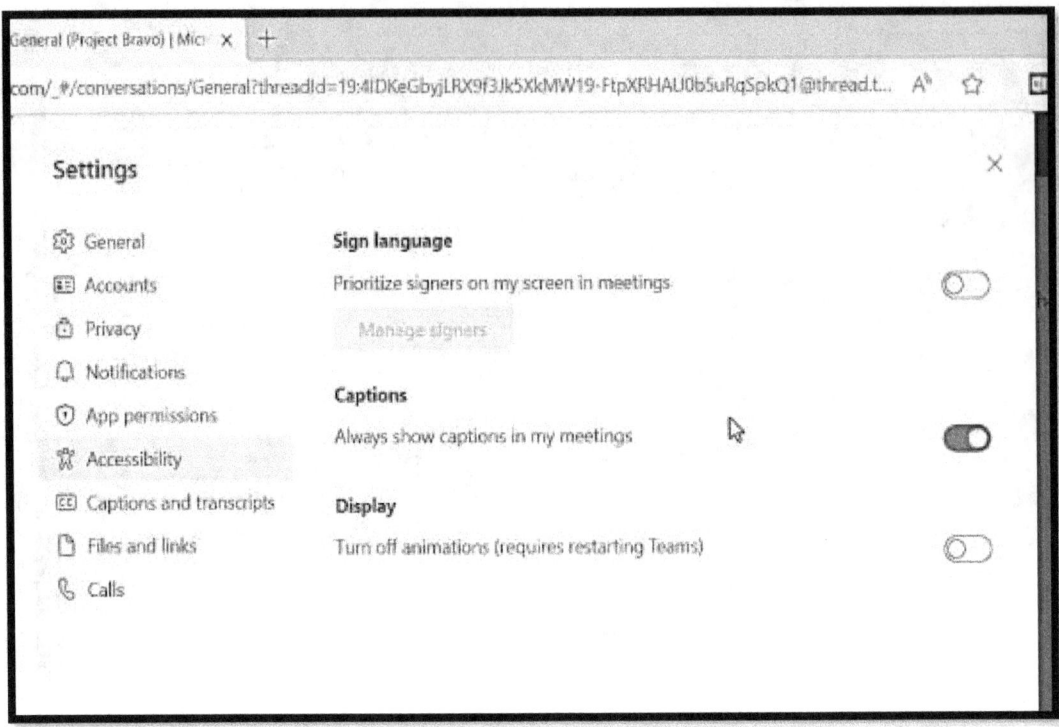

We also have a section for captions and transcripts. You can get Teams to automatically identify you in the meeting captions and transcripts. What we mean by that is that it will pop up your name in the onscreen captions and also in any transcripts of the meeting. We also have the "Filter Profane words" turned on which is also a good idea. Those are a few options that you should check before setting up a Teams meeting.

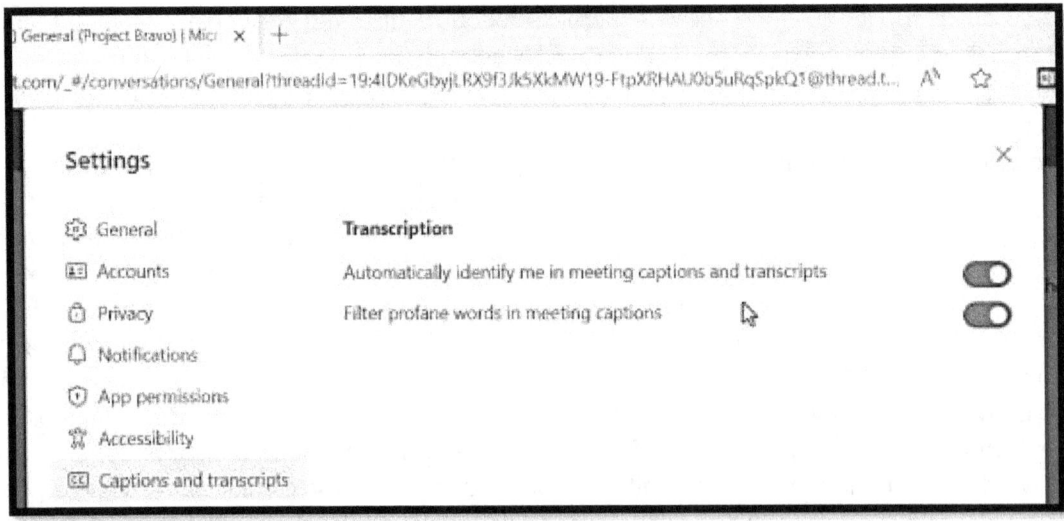

Starting a meeting

The next thing we want to look at before we dive into a live meeting is where you can find all of your meeting options. Let's say you're currently in the general channel having a conversation with different people here and you decide that this would better be discussed in a video meeting. You can start a new meeting from within any channel in Teams. In the top right-hand corner, you have a Meet button, and if you click the drop-down you can choose to meet now or schedule a meeting. The difference between these two is that if you schedule a meeting, that's a meeting that's going to occur in the future whereas meet now will start a meeting immediately.

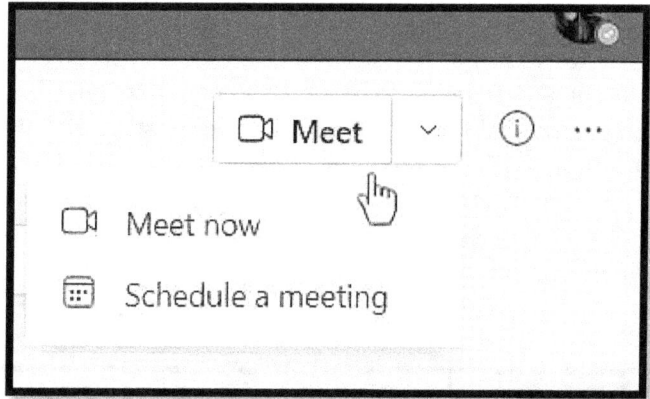

If you go across to the private chat window you can also start new meetings from here as well. In the top right-hand corner, you have a little video call button or you can start an audio call and this will start the call immediately.

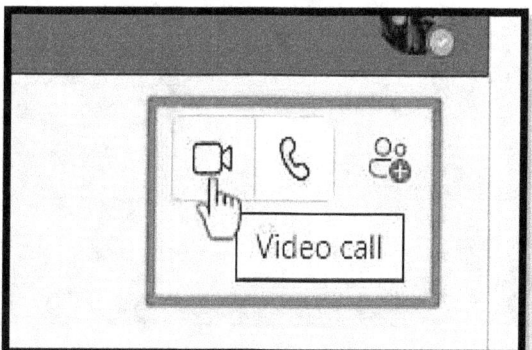

You can also start a meeting from your Team's calendar. If you click on Calendar, you'll notice that in the top right-hand corner, you have a "Meet now" button or you can choose to schedule a meeting underneath here. With these three different areas, you can create or start a new team meeting.

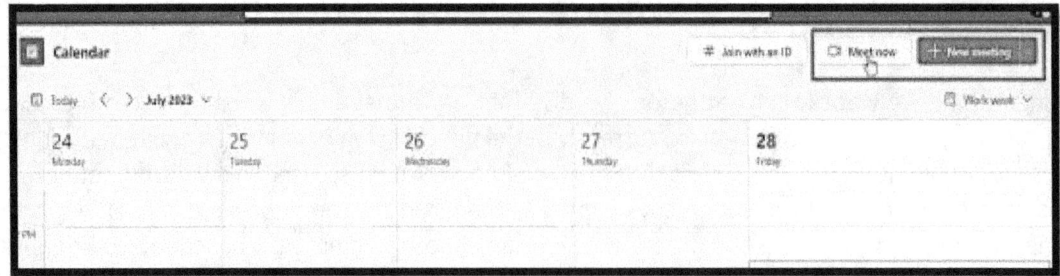

Initiate a call on demand

Now let's take a look at how we can initiate a video or audio call on demand and we're going to start by initiating an on-demand meeting from within a team's channel. Once again within the team, let's say you're on the General Tab, you're having a conversation with Mary and Alex and you decide to jump into a meeting to discuss this in more detail. If you go to the top right-hand corner you can see that you have a Meet button just here with two options: Schedule a meeting and Meet now. "Meet now" is going to take you directly into a meeting and it will take you to the video settings page. This is where you can set a name for your meeting, specify if you want to have your camera turned off or on, add a background filter, and set up your audio.

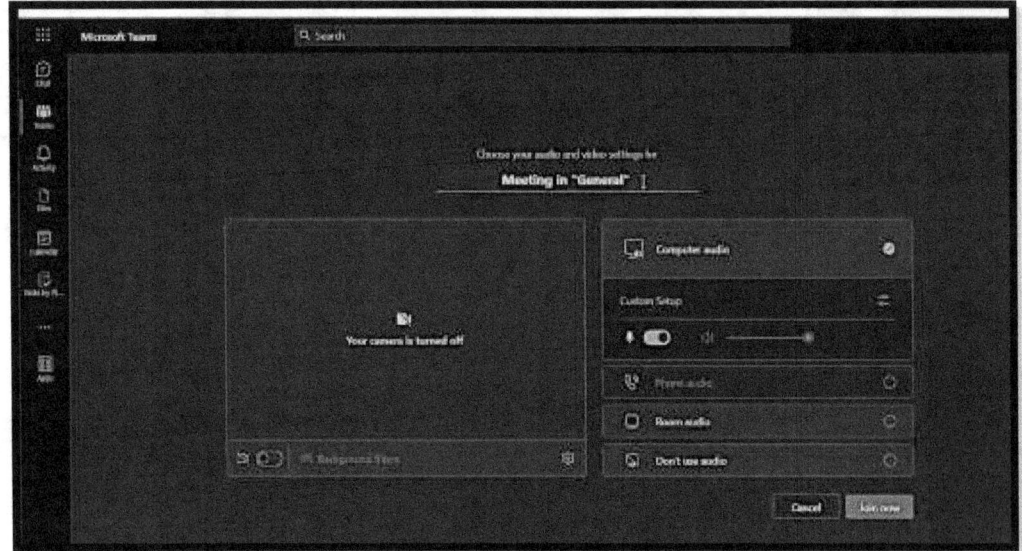

The background filters help disguise your background so if you are working from home and you don't want people to see that your background is a bit untidy you could add one of these background images or you can use your own image. Alternatively, there is a blur option which will just blur out the background as opposed to adding a picture. You can also choose to turn off your camera.

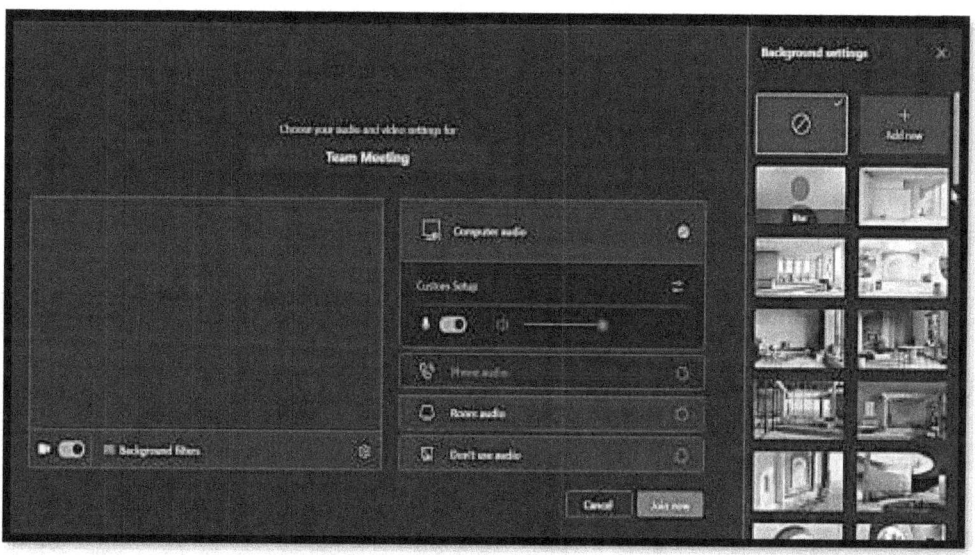

You have a little Settings icon here and this is where you can set up all of our device settings if you need to so just go through and check to make sure that your audio, speakers, and your microphone are set to your correct defaults and this will differ depending on what your setup is like. For instance, you can use a stereo microphone which is the default for microphone use but then your speaker could be set to your laptop. If you had a headset plugged into here, instead of using a standalone microphone you'd be able to select that from the drop-down and use that for your speakers and also for your microphone so just come in here and make sure that your settings are set up correctly.

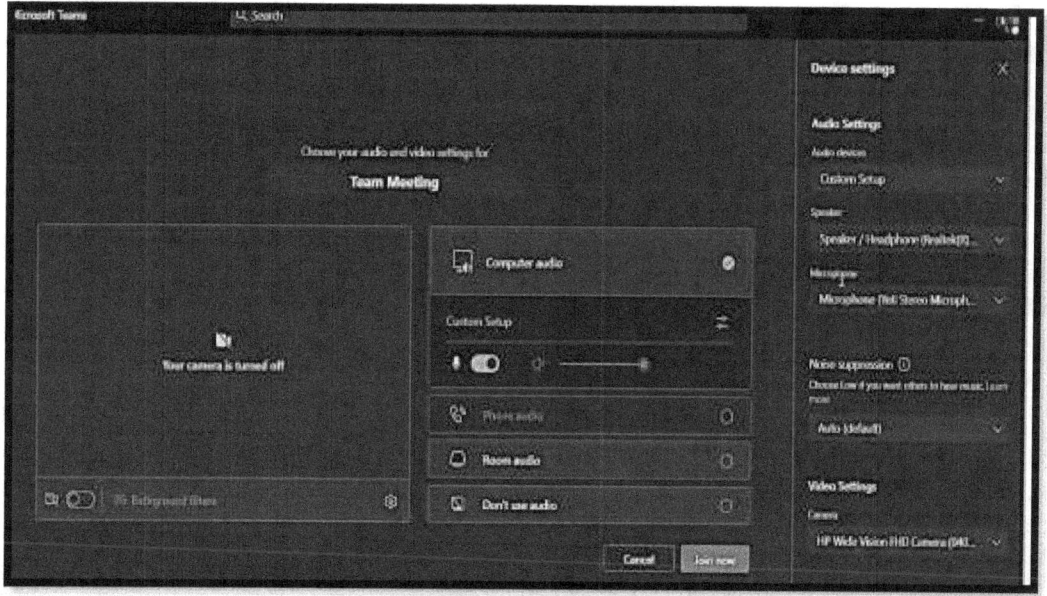

If you do have a webcam attached or you're using a laptop that has an inbuilt camera, the Video settings are where you can select the camera that you're using for your video call. After making these adjustments then you can click on the "Join now" button which is going to take you to the meeting. At this stage, you're going to get the option to invite other people to the meeting so you can copy the meeting link and put that into a team's Channel, add it to an email, or you can add participants directly. Let's say you decide to add a participant. It's going to ask you to confirm the language everyone is speaking.

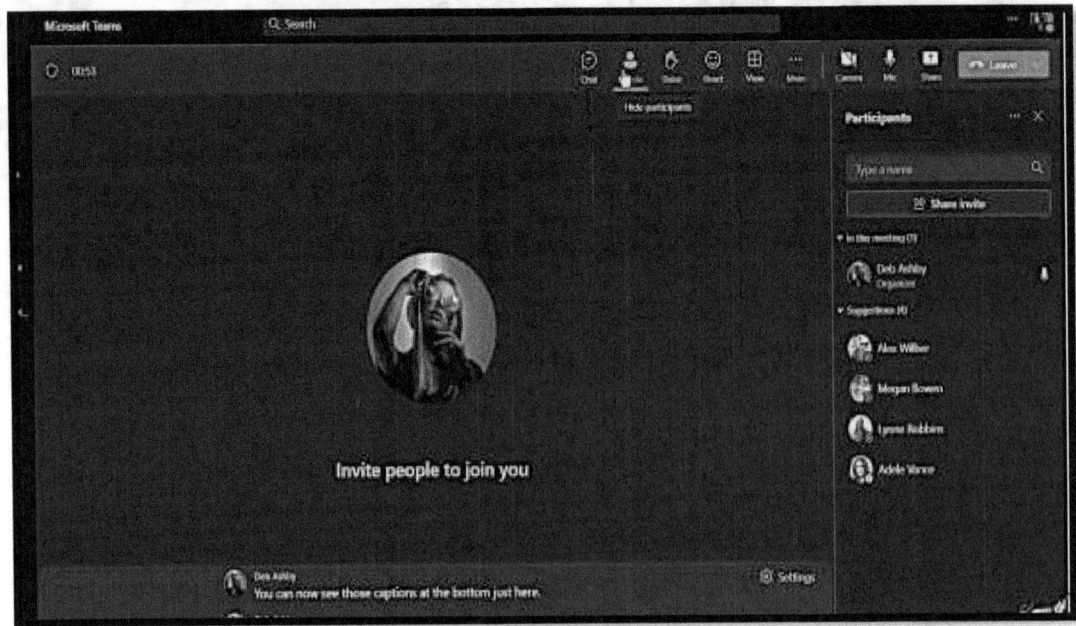

When you're in the meeting, you'll notice a couple of things. Because you turned on captions in the last section you will see those captions at the bottom. You'll also notice that you have a menu bar running across the top with lots of different options related to this team's meeting and because you selected to add participants it automatically popped open the participants' pane for you from which you can go in and can add people to this meeting. You're going to choose "request to join" so it's now calling them and then they can join the meeting. Alternatively, you can type their name at the top and invite them to participate.

The other thing that you have up here is a Chat panel and this is very much like chatting in a team Channel. You can utilize this to have conversations with your team members and this is particularly good because it's a non-audio way of having a conversation. If you're presenting something it means people can have a little chat about what's going on the screen without interrupting the flow of the presentation of the speaker. You'll notice underneath here you have other options as well similar to when you're posting a message. You can attach files, and add emojis, gifs, and stickers.

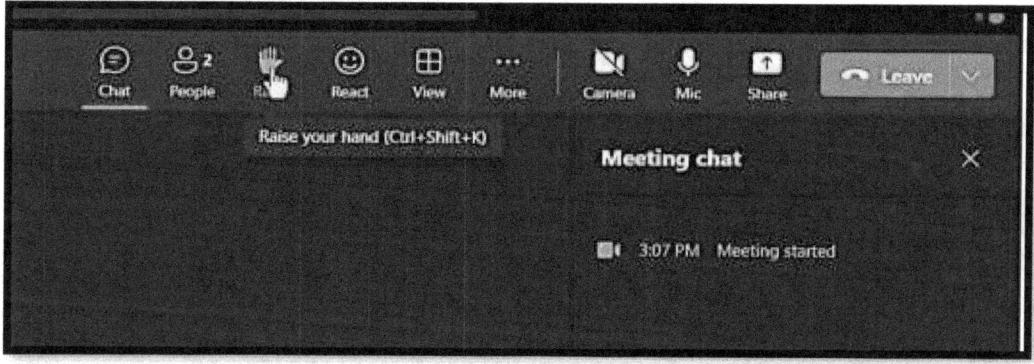

The next option is to raise your hand. The keyboard shortcut there CTRL + SHIFT+ K. Quite often you'll see this in the people pane. When someone raises their hand it is a nonverbal indicator that tells the organizer or presenter of the meeting that somebody wishes to ask a question or make a comment but it doesn't interrupt the flow of the presentation. If you see a hand has been raised you can then turn on their microphone if you have people unmuted and then they can lower the hand once it's been dealt with.

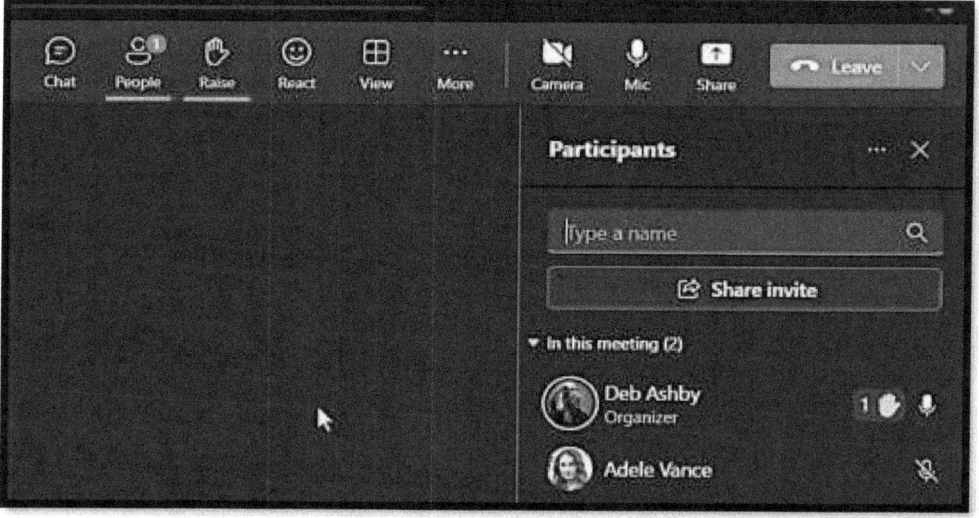

There are a couple of ways that you can lower your hands. You can click on the Raise button again or press CTRL + SHIFT+ K. This shortcut is like a toggle to raise and lower your hand. The next action you can do in a meeting is sending a reaction. If somebody is making a presentation and there's something that you particularly like you could send a little heart and it'll put Hearts across the screen so that they know that that part is appreciated. This is just a nice non-verbal way of interacting. You can change your view so you don't necessarily have to view things in this Gallery format. You can view in speaker format which is slightly different and what you'll see here is if you have more people in this meeting you would see the faces of all the participants in this area.

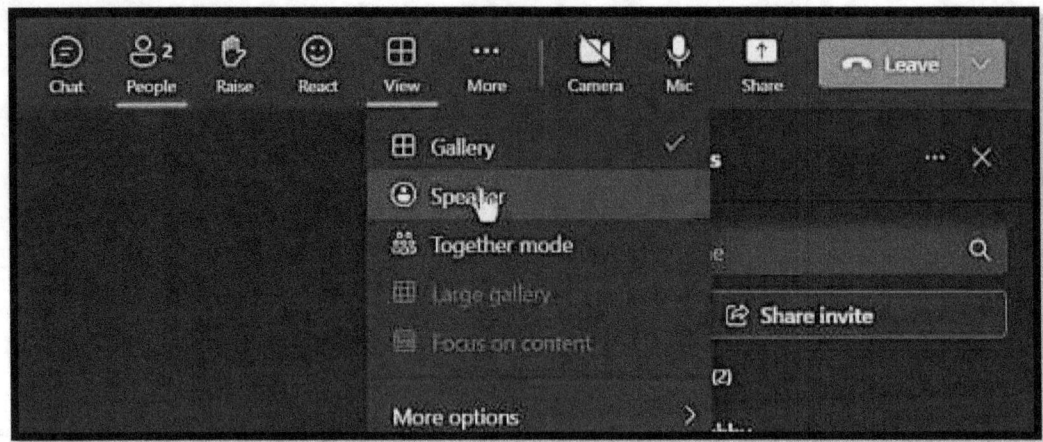

If you go back to change your view you also have together mode and this is a cool little mode that some people weren't entirely sure about when it was first released but which was released during the pandemic to give people more of a feeling of togetherness. If people had their cameras turned on you would see them sitting here in different seats. Some people love this mode while others do not.

You then have the "More actions" button and this is where you can do things like record and transcribe the meeting. Quite often you'll want to record your meeting for a review later or to send to people who can attend; this is where you're going to find the "Start recording" button and as soon as you start recording you'll see that indicated in the top corner - you just have to remember to stop the recording when the meeting has finished.

You have a meeting info area and this is going to give you all of the links to this meeting. You could copy the "Join info" and paste that into an email to send it off to other people. Again, from the More Options area, you can access background effects. You've already seen how this works. You can also customize your language and speech and then you have some further settings that we've already looked at such as Device settings, Meeting options, and Accessibility.

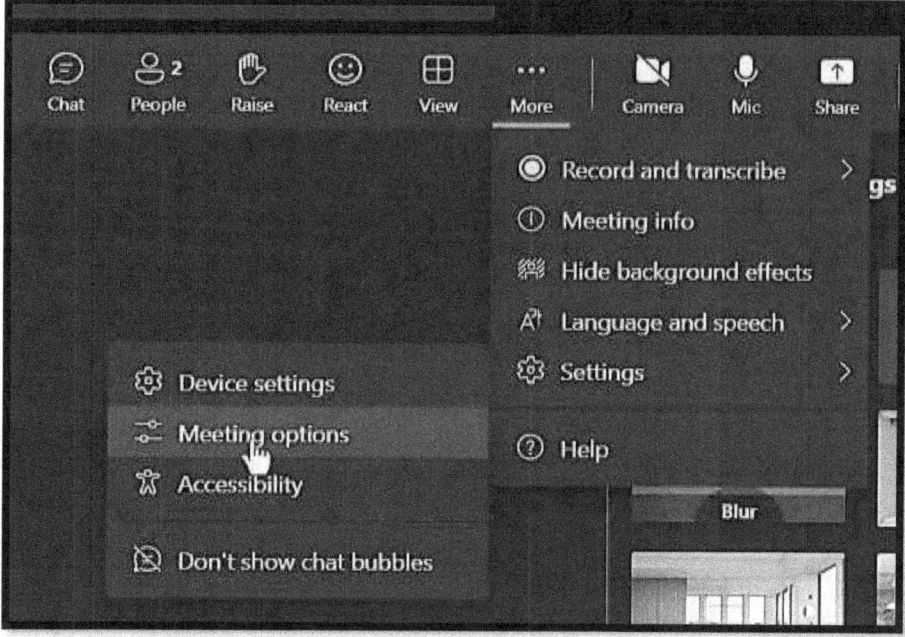

At the top here you have your little camera toggle. You can switch between having your camera off or on throughout the meeting. It's not a case of you turning it on at the beginning and having it stay that way. You can just toggle it using this icon.

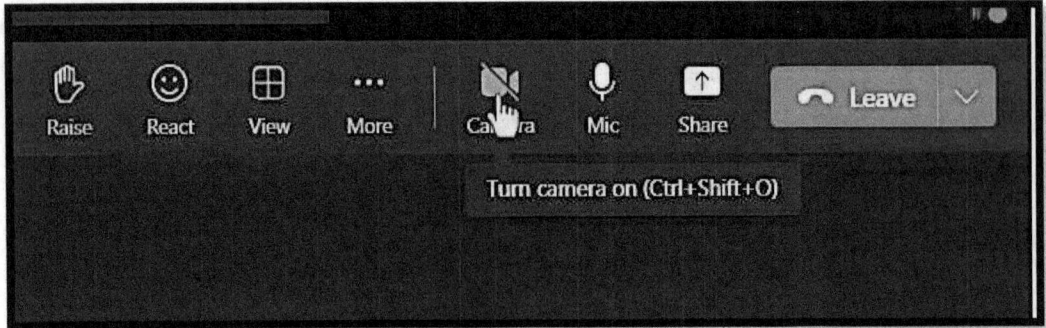

The same thing applies to your microphone. If someone is presenting information it might be that you want to just go on mute whilst they're talking through things and come off mute when it's question time so you can control your own microphone from up here. Then you have the "Share" option at the end and the final button that you have in this meeting is the "Leave" button. Clicking the drop-down brings up two different options: Leave or End meeting.

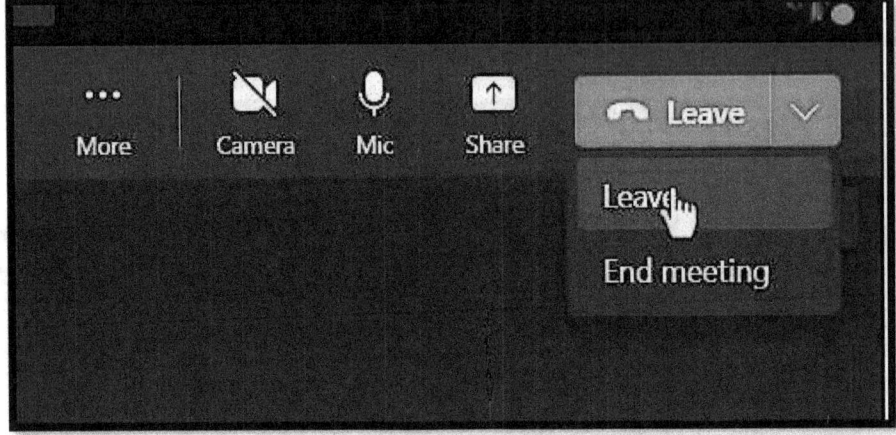

The difference between these two well is this: if you leave the meeting it means that the meeting is still going to carry on with the other participants whereas when you end the meeting it's going to end for everybody in the call. After you end the meeting it's going to close the team's meeting and take you back to the channel and what you'll notice is whichever Channel you started it in you can see here it says the team meeting has ended and the team meeting took 11 minutes and 53 seconds, for instance. You can say the call quality was excellent and send that through and then what you'll find is that if you did select to record the meeting that will be posted in the channel as well so you can access the recording directly from within the channel. Note that if it was a long meeting you might have to wait a little while for that meeting to show in the channel but it will get there eventually. Those are the basics of initiating an on-demand meeting using the "Meet now" button.

Schedule a meeting

In the previous section you saw how you can start an on-demand meeting using the Meet Now button and in this section, we're going to take a look at scheduled meetings and we're going to show you how you can schedule a meeting from within Teams and also from within Outlook in Microsoft 365. We're going to stick with the General Channel and once again you'll go up to the "Meet" button in the corner but instead of clicking on "Meet" you're going to click the drop-down and choose "Schedule a meeting." You'll notice that this pops open a calendar invite, similar to creating an event or an appointment in your Outlook calendar.

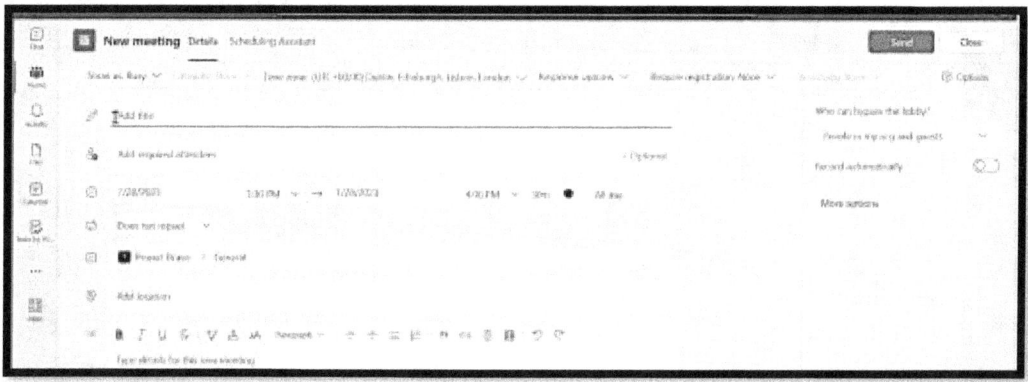

You can add the title for the meeting and add the required attendees. Something that can help you when it comes to scheduling is to utilize the scheduling assistant tab just here. If you click on Scheduling Assistant it's going to show all of the attendees of the meeting (you can see them all listed) and it just helps you pick an available slot where everybody is free. Then you can set a start date or a start time and if you want to check before you send this that everyone is available, you can go to the Scheduling Assistant and you will see if everybody is free in their calendar. That's a great little tool to help you with your scheduling.

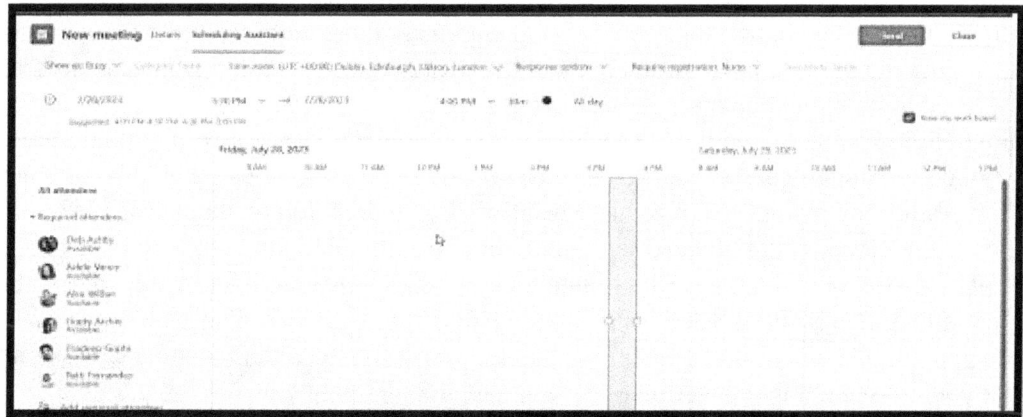

You also have the option to choose if this repeats let's say a regular or weekly team meeting you could choose to set this to repeat on certain days of the week. You can see the channel that this is posted in and you could add a location and a description. Notice on the right-hand side you have a little options button, once again you can come here to set up who can bypass the lobby. You could set this so that everybody who's invited to this meeting can bypass the lobby. You can also choose to record this meeting automatically and that can be useful because you may lose track of time when you start a meeting and forget to start the recorder so set it up beforehand so you don't have to worry about it.

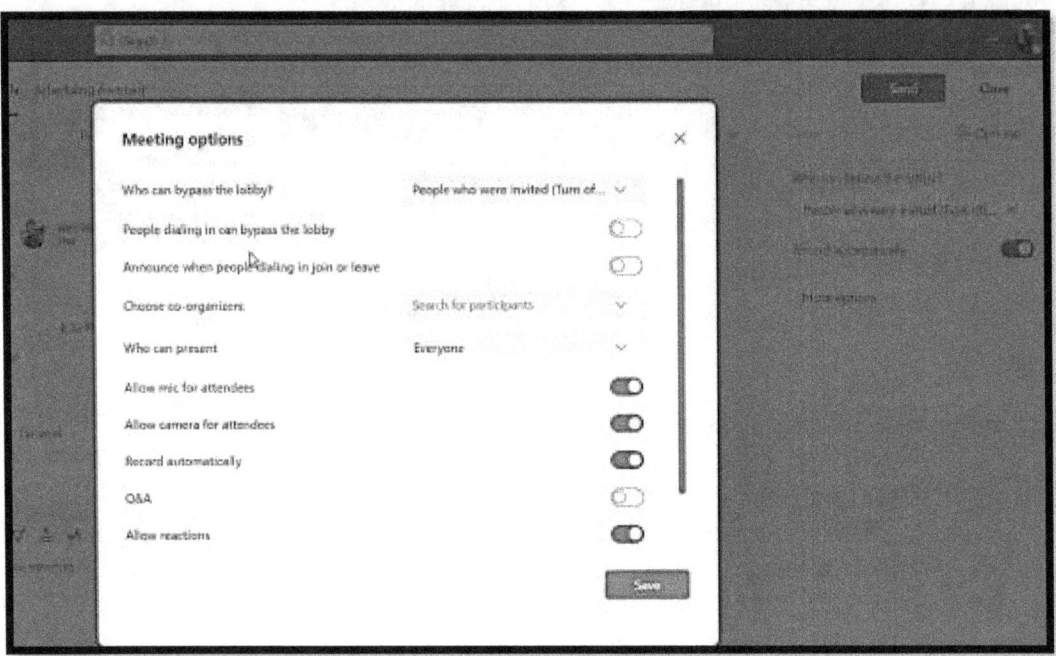

You'll notice you have more options and if you click this, again, you can customize your settings further so if you want people to have their microphones turned on when they enter the meeting make sure you have this option toggled on. If you want everybody to be on mute you can toggle that off so just check these before you send out the invite. You also have some options here for registration. If this is a meeting that you want people to register for you could set that up from this "Require registration" area and this means they need to register usually with their email address which can be useful particularly if it's going to be quite a large meeting because it's important to know roughly how many people are going to be attending so don't forget about that option. After you're done with the settings and you're pretty happy with this, click on "Send" to send that out. Immediately, what you'll see is that this meeting gets posted in the team's channel so everybody who is a member of this channel can see that sitting just here. The other thing it's going to do is it's going to post this in your team calendar so if you go over to the left-hand menu and click on the calendar you can see the meeting is there.

Joining a scheduled meeting

Now as you can imagine you have a couple of ways that you can join this meeting when it begins. If you click on the meeting in the calendar you have a join button just here. Alternatively, if you're working within the Teams' Channel you can click on the meeting here and you have a "Join" button in the top corner. When you do reopen a meeting that you've sent you'll see that a few details have changed. For example, if you take a look at the details tab you can see that you now have access to some tracking information so you can see here who has accepted and who has rejected this meeting request.

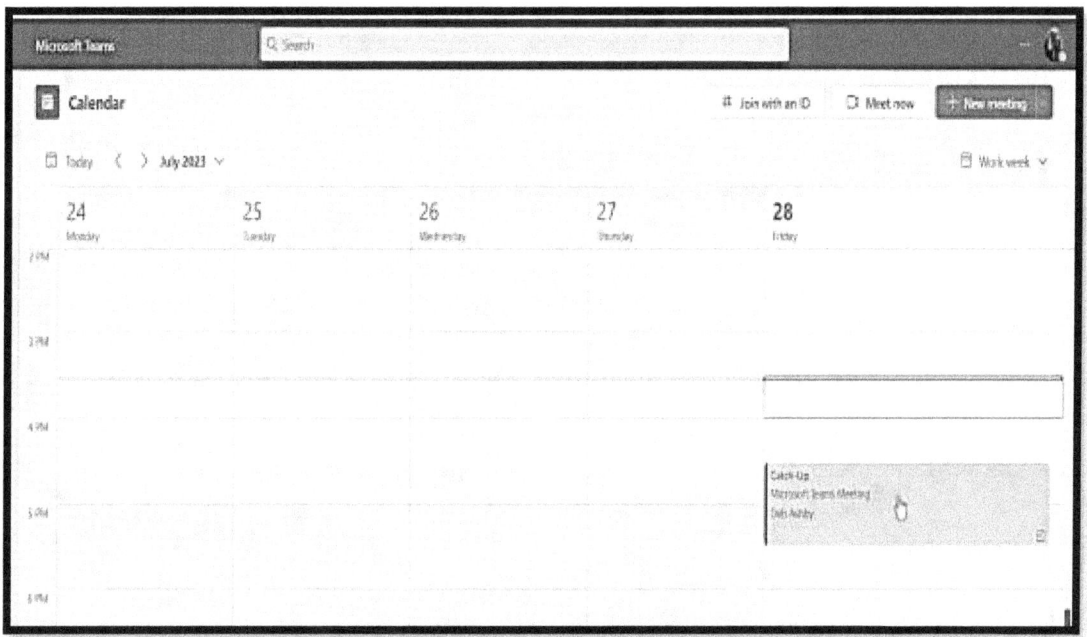

What you'll also notice is that in the body of the email you now have the meeting information and this information contains links so that people can use these links to join the meeting. You also have some additional tabs at the top ready for when the meeting is completed, for example, you have an attendance tab that currently doesn't have anything in it because the meeting isn't over but once it is you're going to see a list of all of the people who attended the meeting. You also have a meeting whiteboard Tab and a Q&A Tab. You can see it's very simple to schedule a meeting and then join that scheduled meeting.

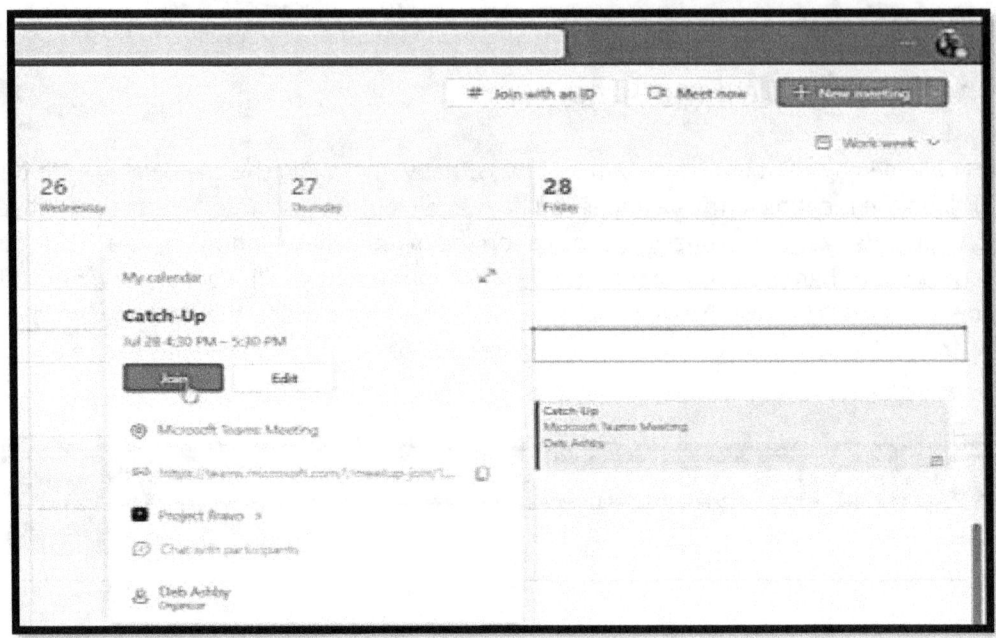

It's worth noting that you also have options to join Teams meetings from within Outlook. If you go over to Outlook and go to your calendar within Outlook you can see that this also reflects that team meeting that's been added to the calendar. Double-click to open this up and you'll see that meeting information. You can join the meeting either by clicking on the link in the body of the message or you can click "Join Teams Meeting" at the top.

Another thing worth noting is you can also create Teams meetings from within Outlook. If you go to Monday, next week, for instance, and double-click to create a new event you can create a Teams meeting or turn this into a Teams meeting by simply clicking on the Teams meeting button here. There are lots of different options there when it comes to scheduling meetings, ensure you have a play around with them.

Sharing your screen

In this section, we're going to take a look at how you can share your screen and files with other team members. Let's say you're in a meeting and you want to share different things with your meeting participants, you'll notice you have a "Share" button just here. The keyboard shortcut to open this is CTRL + SHIFT + E. When you click on share, there are many different things that you can share and we're going to focus on sharing a screen, a window, or a tab.

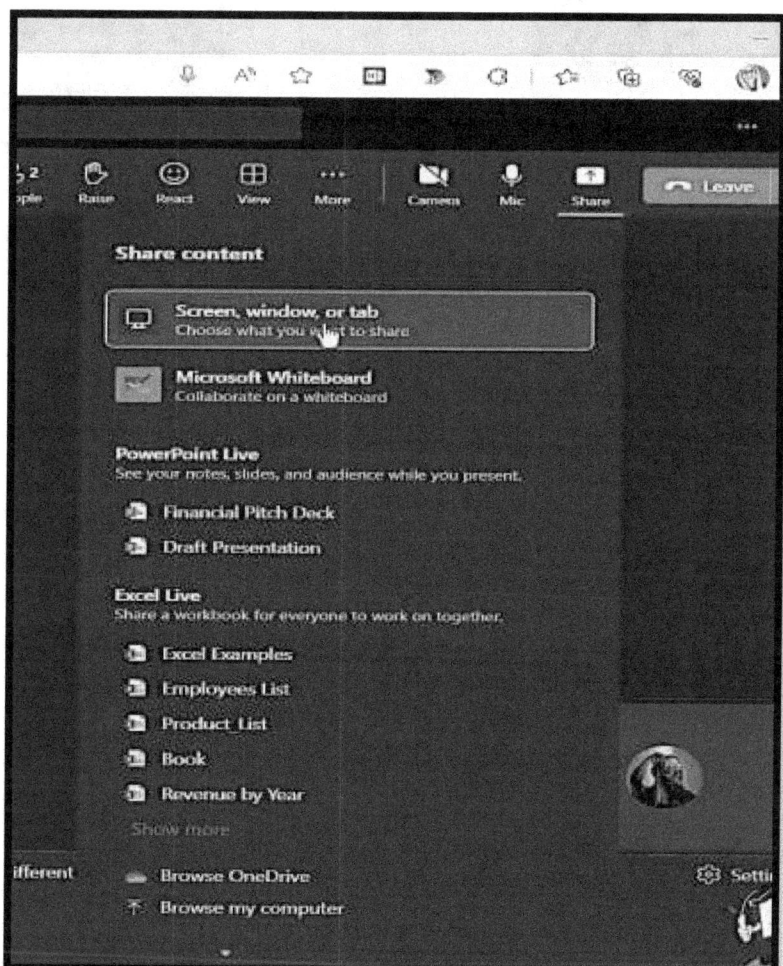

When you click on this option it's going to pop open a little window that allows you to select a tab in Microsoft Edge, it could be an application or you can simply share your entire screen. The difference between all of these and what people will see is this:

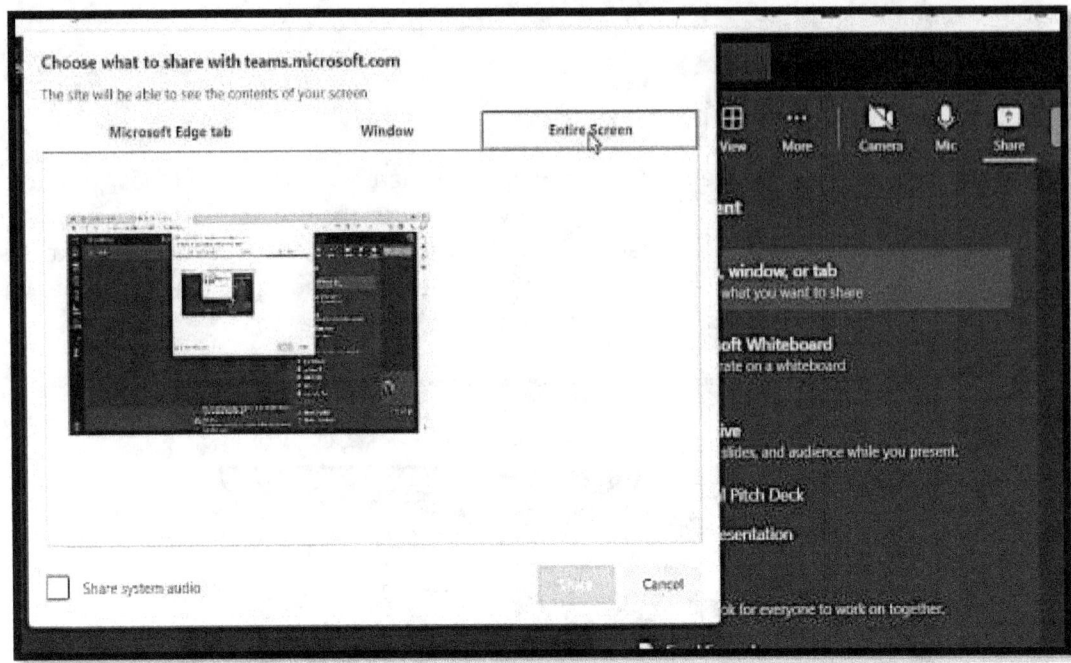

- The Microsoft Edge tab is related to if you're using the Microsoft Edge browser. If you have a web page open specifically in Microsoft Edge you could choose just to share that tab.
- If you choose Windows, this is going to show you little thumbnails of everything that you have open on your desktop so if you have your Outlook calendar open and you want to do a demonstration in Outlook, for example, you could choose to share the Outlook window with your team members.
- The final option is the entire screen so this will display whatever you're looking at currently on your desktop.

Now the difference between choosing a window to share and choosing to share your entire screen is that when you share your entire screen it means the participants in the meeting are going to see everything that occurs on your desktop so if you have your Outlook showing they will see all of your emails, if you have notifications coming in everybody's going to see those as well. Some people don't necessarily like other people to see that type of thing and so prefer to share just a window as opposed to the entire screen. If you were to share your Outlook window, selecting it and clicking on "Share" means it's locked to this window so it doesn't matter if you have pop-ups coming up on your screen, it doesn't matter if you click around and open up other things, people are only going to ever see the application or the window that you chose to share. You could be clicking around on lots of different other things and they're not going to see you doing that. You'll notice that when you are in Sharing mode, right at the bottom you have a hide button so you can hide this little panel and when you want to stop sharing you can come back to this window and click on the "Stop sharing" button it's going to take you back into the meeting and out of sharing mode.

One thing to note here while we're on this screen is to check out what you have across the top. It says your status is set to Do Not Disturb. You'll only get notifications for urgent messages from your priority contacts so that's worth noting. Also, if you look at your profile picture you can see you have the little red icon over the top so because you're in a meeting and you're presenting Teams has set it to do not disturb which will stop notifications coming in.

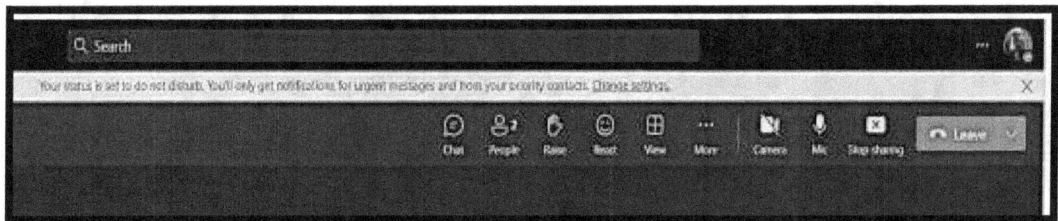

The other thing you can share here are files. You have a section here for PowerPoint Live and Excel Live and if you simply want to share a file with everybody, you'll notice at the bottom here you have the options to browse OneDrive or browse your computer so you can click on browse your computer and choose a file to share and now everybody can see the file that you've shared. You'll also notice it tells you at the bottom that you're presenting and at the top you have a "Stop sharing" link so that is how simple it is to share a window or share your entire screen or share any file that you have stored off in File Explorer.

Making live presentations in meetings

One of the newer updates to Teams has been the ability to work on and share PowerPoint and Excel files live with meeting participants. If you go across to the Share button, at the top you have two different sections: PowerPoint Live and Excel Live.

Let's start with PowerPoint first. Let's say you want to share a PowerPoint presentation with all meeting participants, you just need to click it, it'll go away and prepare the slides and then it's going to load the presentation up into the window. Now because you're the presenter you'll currently be seeing the presenter View and that means you get to see the presentation, all of the different PowerPoint slides underneath, and any notes that you've added. Note that anybody else in this meeting doesn't see this view, what they see is what is highlighted in red.

The cool thing about working in **PowerPoint Live** is that you have lots of different ways to collaborate with other meeting participants because when you share in this way, you're effectively co-authoring with other people in this meeting. You'll notice underneath you have things like annotation tools so you could grab the pen here and start to draw things on this presentation and the participants will see live any changes that you're making. Now if a participant wants to make their own changes to this presentation they're going to have a "Take control" button on her screen and with that, they can now effectively take control over from you and move through the slide. They can make annotations or choose to stop presenting or if you want to just take back control you can do the same with the "Take control" button on the toolbar which passes the control back to you.

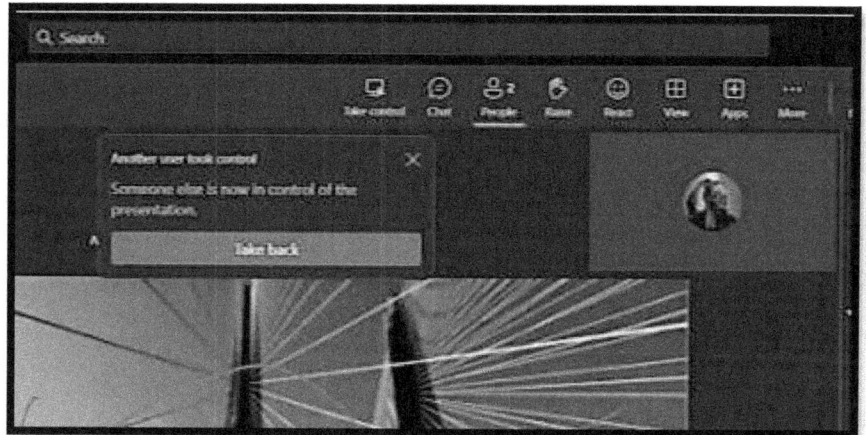

A couple of other things that you have access to is a grid view that allows you to display these slides in a slightly different way. You just need to double-click on a slide to pull it back to the original view and if you click on the three dots you can choose to hide the presenter view so you're seeing exactly what your participants are seeing. You can view the slides in high contrast or even translate the slides into different languages. Another thing that's cool about this is that a participant doesn't have to move through the presentation at the same speed that you are. If you're talking through a slide and Mary wants to go back and read the previous slide she can do that and it doesn't affect what's being presented to everybody else so everybody else is going to see the slide that you're currently on but she has control to move through the different slides and review them without it affecting anything. Let's take a look at **Excel Live**. Once again, if you go to the Share button you have Excel live just here and this is going to load up the Excel spreadsheet and share it with all meeting participants.

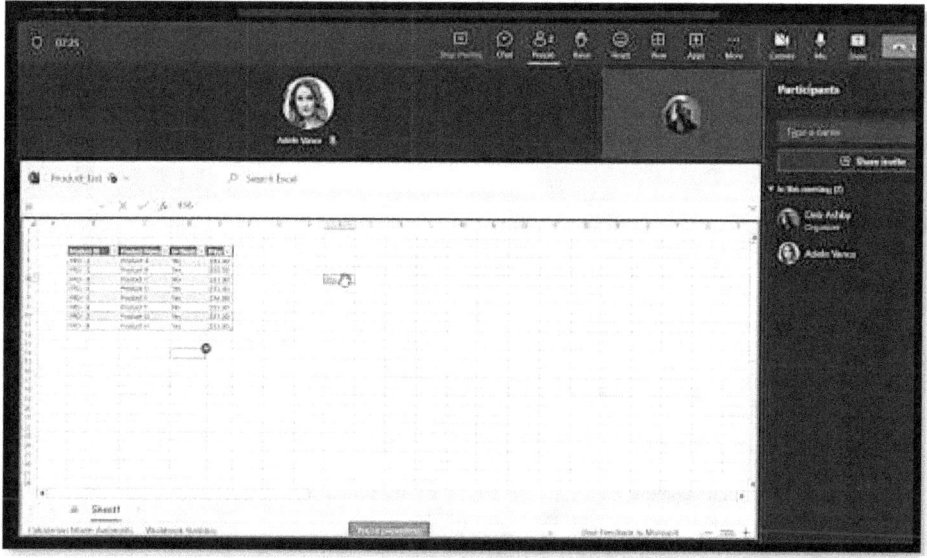

This is cool because this is pure co-authoring. Your participants can see where you're clicking around on this spreadsheet but you can also at the same time see where other participants are currently clicked in. You could type in some numbers in a cell but at the same time Mary could be typing things into the spreadsheet as well and you're going to see it all update in real time. This is the same as co-authoring documents in Microsoft 365. Once you're done, you can click on "Stop sharing" and that is going to close that file down. PowerPoint Live and Excel Live are two great additions when it comes to collaborating on spreadsheets and presentations with other meeting participants.

Brainstorming ideas in a meeting

In the last section, you saw how you can share files and also share a desktop or an application when working in a team meeting. Now if you recall there was another option there and that was sharing a whiteboard with all team participants. What exactly is a whiteboard and how is it useful? The first thing that you need to know is that the Whiteboard is a separate application in Microsoft 365 itself so if we jump back to our Microsoft 365 homepage and click on the app launcher, at the bottom of apps we have Whiteboard. If we click to open this up and take a look at it, what you'll see is that a whiteboard is a blank canvas that helps you brainstorms ideas and projects with colleagues and team members.

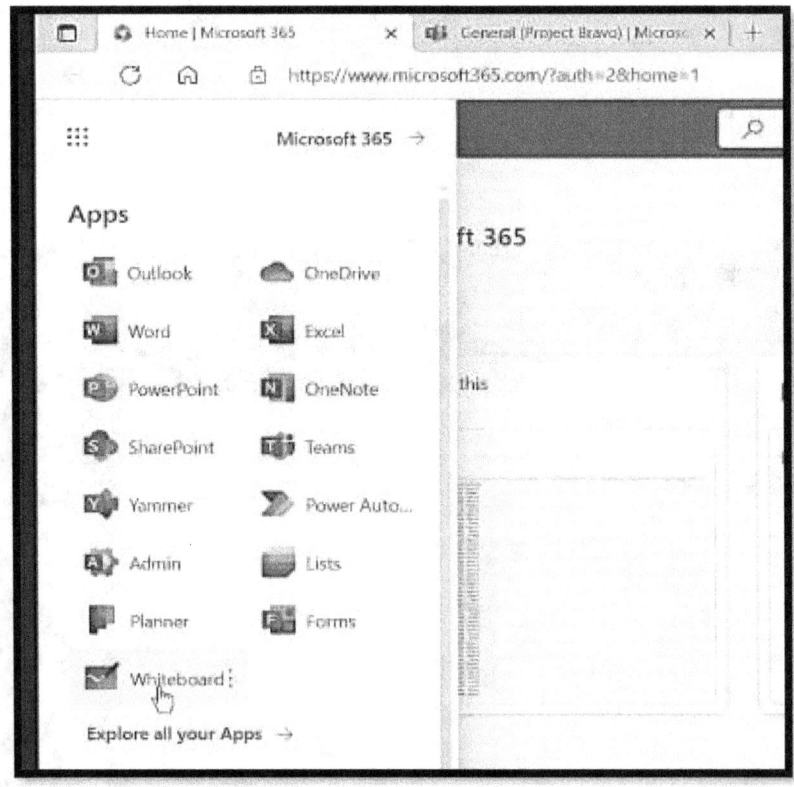

You can create your own whiteboards from scratch or utilize templates to give you a head start and at the bottom you'll see you have a little toolbar that allows you to annotate and add other elements to this whiteboard. For example, you can add little sticky notes so if you select one of these and then click somewhere on the Whiteboard it's going to give you a sticky note-style graphic. You can stick this wherever you like and start typing in some notes.

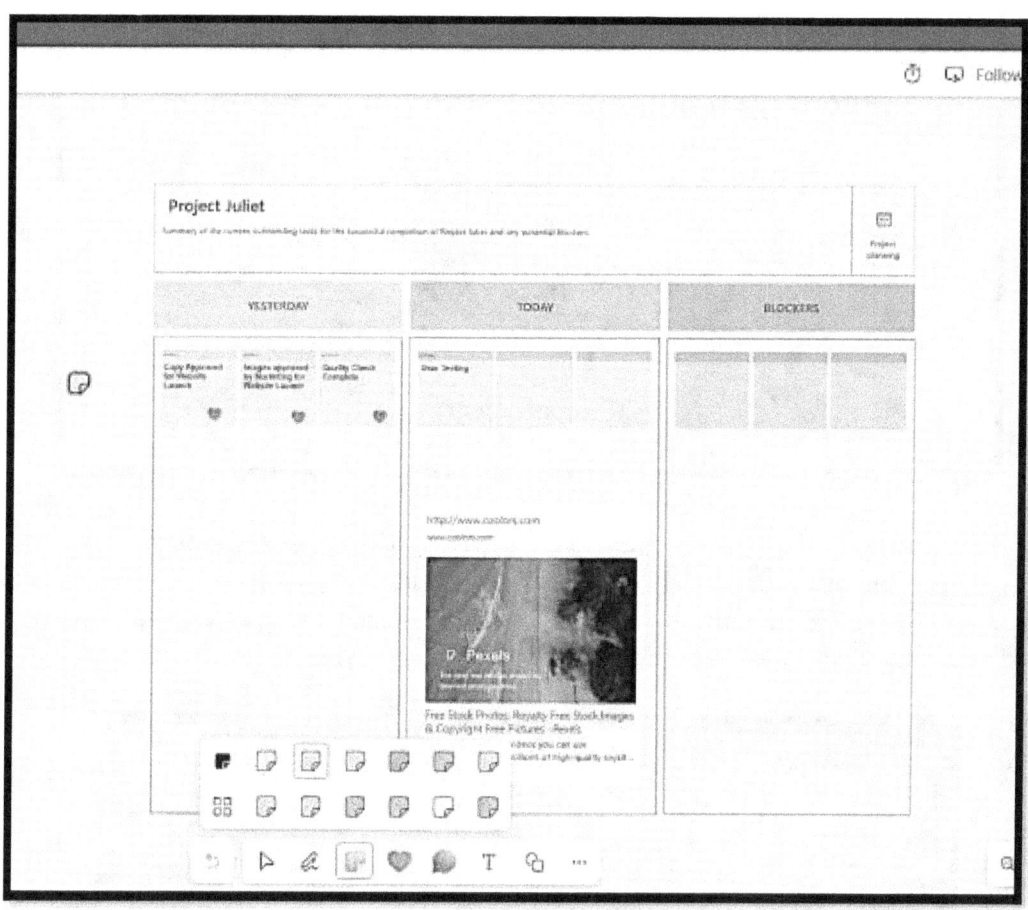

Also, note that when you add text here you have a little floating formatting bar just above which will allow you to change the color of the note. You can delete the note, add comments, or do some further formatting of the text. You can add reactions so if you click on the heart just here you can see you have a few things so let's say you want to click the check mark, you can just put that wherever you like on the Whiteboard. You can add things like comments so you can simply click, and add your comment much like you would if you are working in any Microsoft application and you can see that comment is represented there by the little comment icon which you can move around and position it wherever you like. You can of course add text anywhere on your whiteboard, simply click and start to type and if this is a bit too small you can use your formatting to make it a lot larger. You can also add text boxes wherever you like on your whiteboard so you can just click and then start typing. You can add shapes too so if you want to add a triangle you can do that and then of course, you can drag the corners to resize it. This gives you an idea as to the kinds of things that you could build here. You could make diagrams, and project plans, have a blank canvas, and allow people to write notes and brainstorm different ideas.

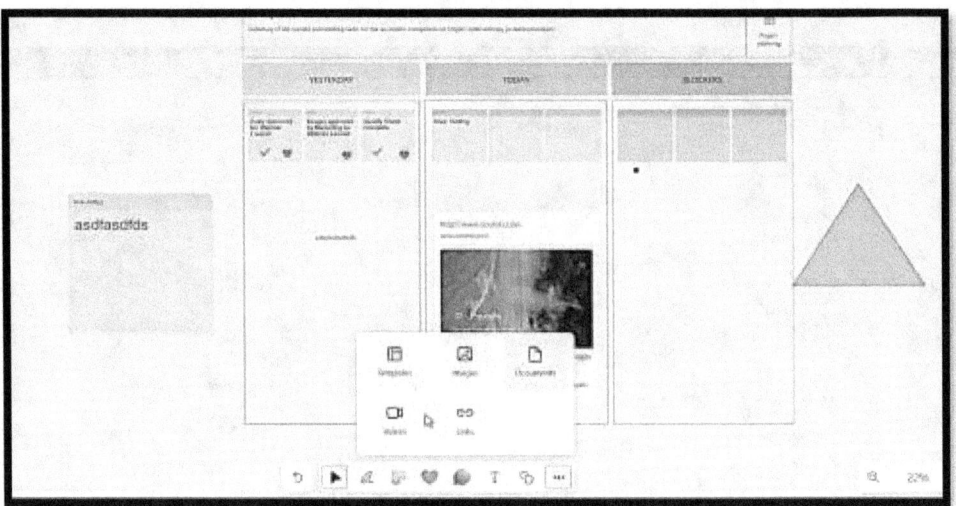

If you click the three dots you have more options. You have templates, images, documents, videos, and links. Again, these are all things that you can add to your whiteboard so you can build up your whiteboard however you want to build it and once you're happy with it; you can rename your whiteboard by clicking at the top here and typing in a name for your board.

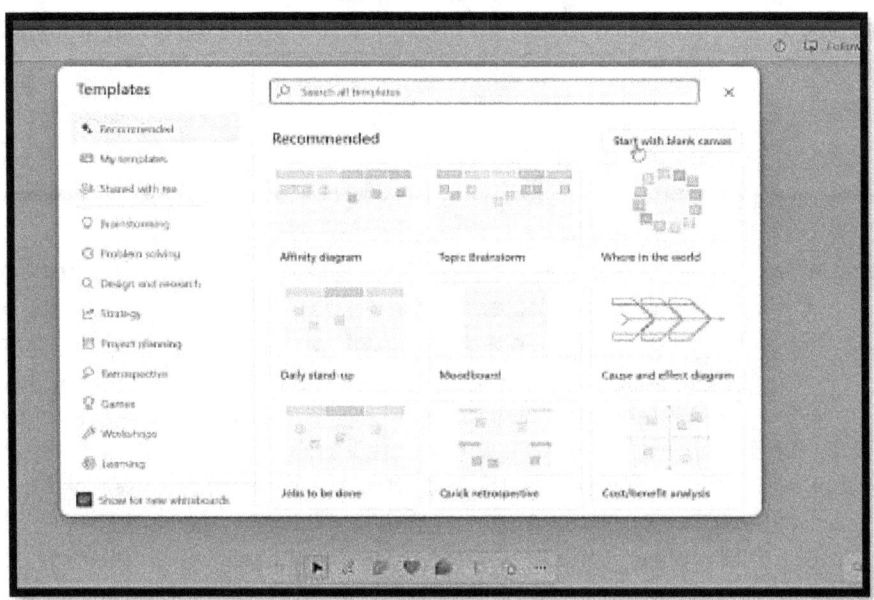

If you want to create your own new whiteboard from here you have the "New Whiteboard" button and you can start with an entirely blank canvas or you can choose one of these templates which are divided into different categories. You can preview it to see what it looks like and if that's something you like, you'd choose this template.

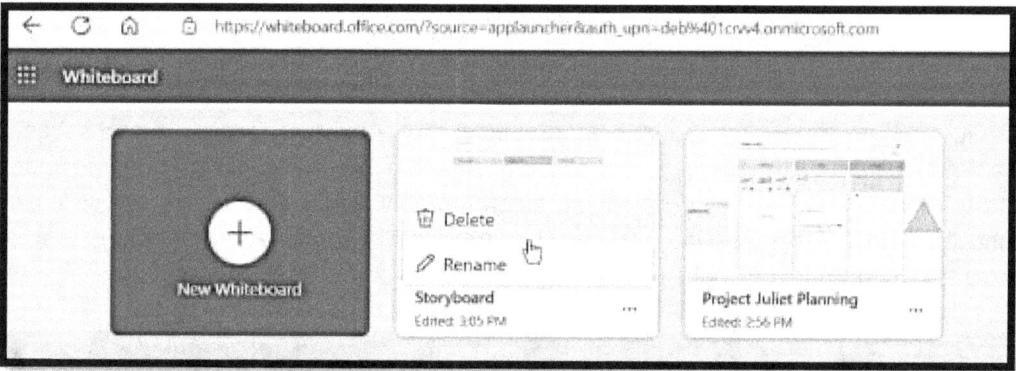

Then you can choose where you want to place it, you can move it around, resize it, and start to customize it so it's suitable for your team. When you're working within the Whiteboard app you have three dots which is going to give you access to more options and there are a couple of things you can do here. You can delete or rename that. Another thing that's worth noting if you are working within one of these whiteboards is you also have a Share button in the top corner so you can get the link to this whiteboard and share it with other people and they can then come in and start making changes to it.

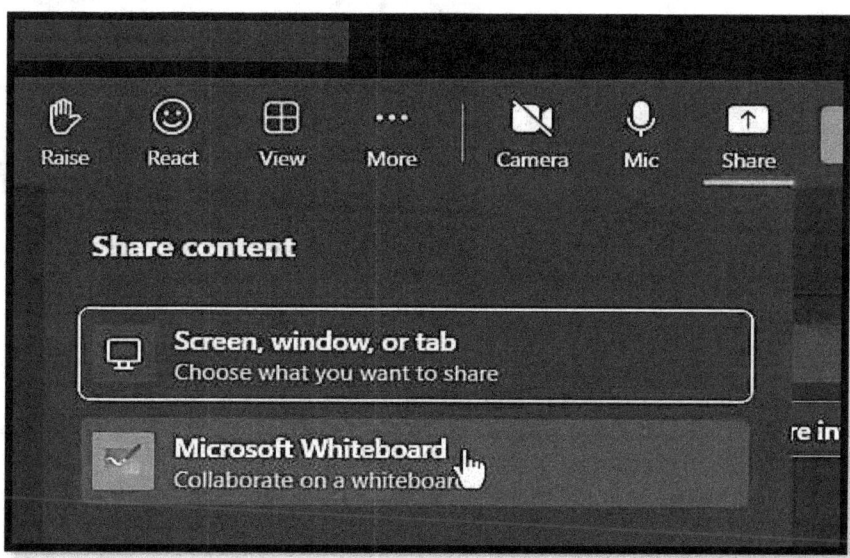

How does this all relate to Microsoft Teams? Think about it this way: when you're in Microsoft Teams you could choose to just create a new whiteboard when you're in the meeting and that whiteboard will give your team participants a blank canvas where you can brainstorm ideas and everybody can dive in and co-author the document together so you can make changes, your colleagues can make changes and you'll see those updates on the whiteboard in real-time. Alternatively, you could set up a whiteboard in advance of the meeting and then share that whiteboard with meeting participants. Let's say you're already in a meeting, what you could do is share that whiteboard and work on it with your participants. If you go to the Share button in the top corner you can see you have an option for Microsoft Whiteboard. Click on that to load the app and you will see all of those existing whiteboards so if you want to work on a particular Whiteboard you can just click to share that and you can now work on it together. There are times when you share a Whiteboard with other meeting participants that you may not want other people to be able to edit the Whiteboard. You can turn that setting off for individual whiteboards when you're in the meeting. Click on the settings icon at the top, you'll notice you have it set here that other participants can edit but you could just toggle this off to stop editing other than yourself.

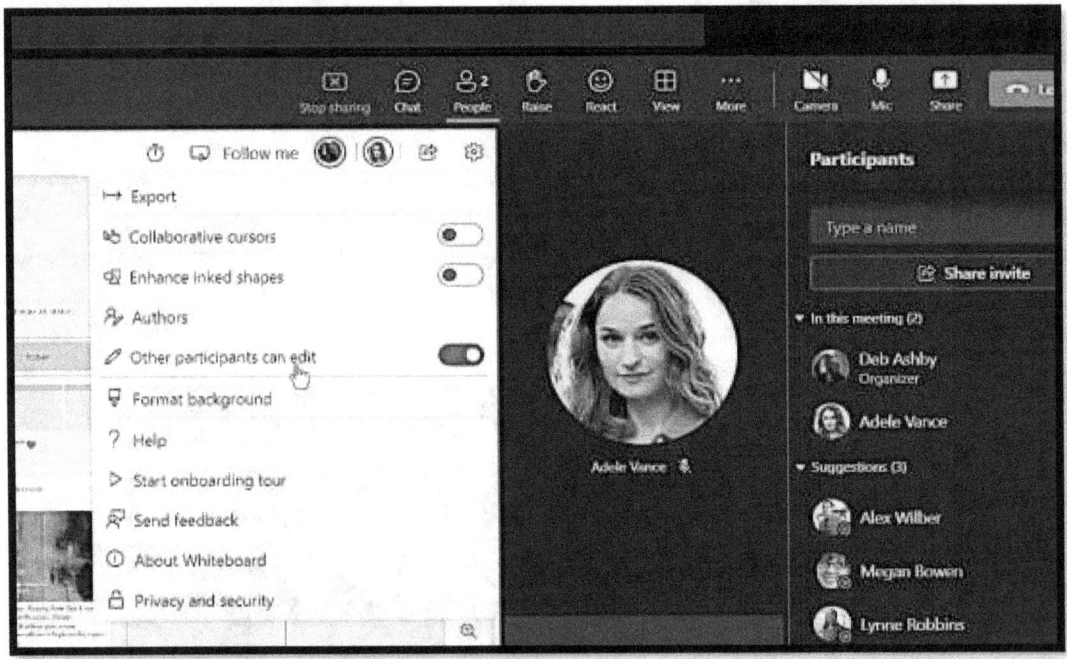

Once you're done working on the Whiteboard you can click on "Stop sharing" and that will save those changes back to the Whiteboard app. Don't forget if you want to start a new whiteboard from scratch to brainstorm ideas again, you just go back into Share a Microsoft whiteboard and instead of selecting one of the existing whiteboards that are there, you can simply click on the "New Whiteboard" button and there you have a blank canvas to share your ideas.

Advanced meeting scenarios

Breakout rooms are a great way for a meeting organizer to manage participants into different groups in a large meeting. For example, you might have 20 people on a call and the meeting organizer wants to send you off into groups of four to work on a question that they've posed or have a group discussion. Imagine breakout rooms like having many workshops going on in four different meeting rooms and it does simulate the physical collaborative process. Let's take a look at how Breakout Rooms work. Note that we've been using the Microsoft Teams web app accessible through the Microsoft 365 portal and that's fine. You can do pretty much everything using the web app but you cannot access Breakout Rooms so you will need to switch across to the desktop version of Teams if this is a feature that you'd like to use. It's also worth noting that Breakout Rooms is one of the features that are only available in the business subscriptions of Microsoft 365. If you've switched over to the desktop application, you'll notice you now have this Room's icon at the top. If you open this up you can invite participants because normally for a Breakout Room, you want to have lots of people in your meeting.

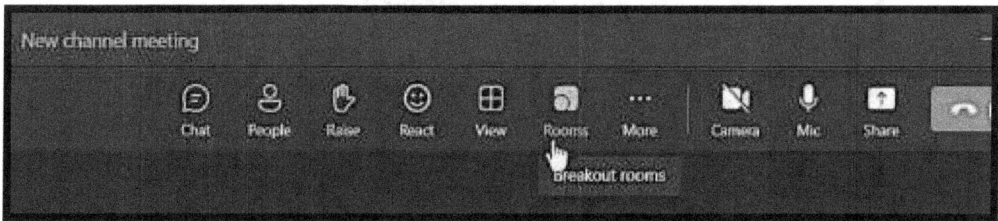

Let's say that as the organizer you want to create some Breakout Rooms. Click on the Rooms option and you can see a pane on the right-hand side. Now you get to choose the number of Breakout Rooms that you want to create and from the list, you can create up to 50 breakout rooms. We wouldn't necessarily suggest that because that is going to be hard for you to manage as the organizer but if you have a few people on you might want to create two to four rooms.

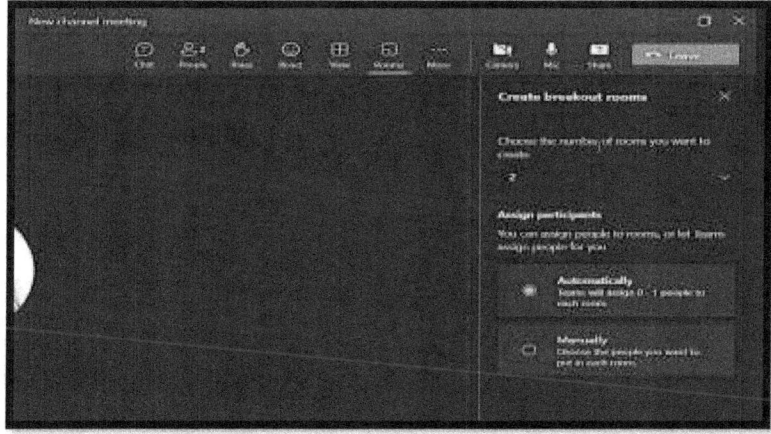

Let's say you chose two rooms, you then get options for assigning participants to those breakout rooms and you have two options:

- If you leave it on "Automatically" Teams will do all of the assigning for you. It will equally assign participants across those two rooms in a very random way.
- If you want a little bit more control and you want specific people to go into specific teams or specific people to work with other people then you would choose "Manually."

Let's go with that second option then you're going to say "Create rooms" and you can see it's now created two rooms that are currently showing as closed with a zero after them because you don't have any people assigned yet. If you have people here you would see all of their avatars across the top and you would go through and you would manually assign them to Rooms. Also note that under "Assign participants" you have some icons:

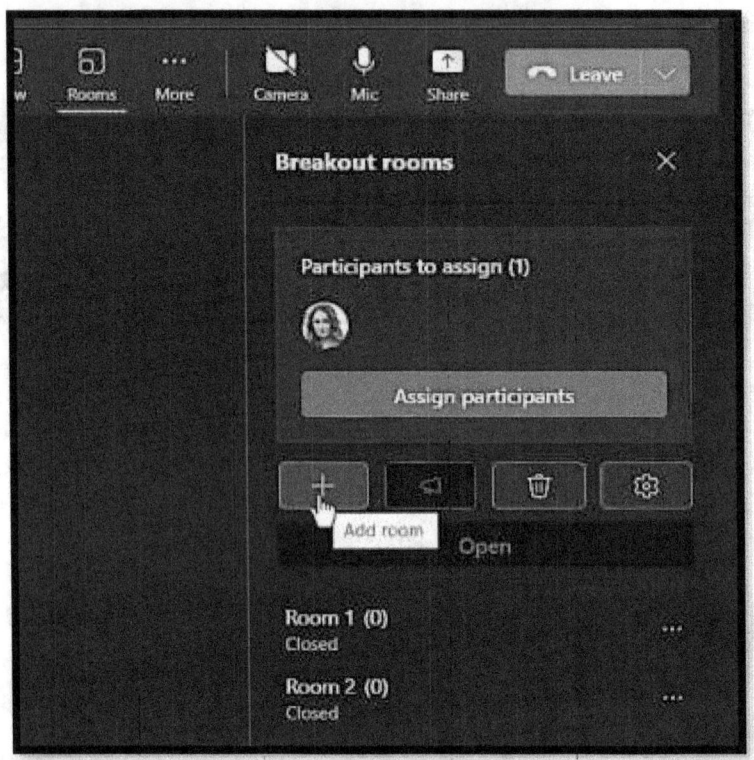

- You can create new rooms on the fly so if you decide that you want three rooms or four rooms you can add one or two in.
- You can also make an announcement. This option here means that as the meeting organizer you can essentially make an announcement to everybody in every single room.
- You can delete rooms.

- You also have some settings for our rooms as well. You can do things like assign presenters to manage the rooms and set a time limit for people to be in those rooms, so for example you could set the time limit to 10 and then after 10 minutes it will automatically close out those breakout rooms and bring everybody back to the main meeting.

After making the necessary settings, click on Save and you can see now that you have the rooms with everyone assigned. You can set this up before the meeting begins so that when it comes time to do the exercise or whatever it is you can simply go into rooms and click on open and it will open those Breakout Rooms and assign everybody off into those separate rooms. The participants disappear off the main meeting window because they're now effectively off in their own rooms and of course, those participants will just see all of the other people that have been assigned to that room.

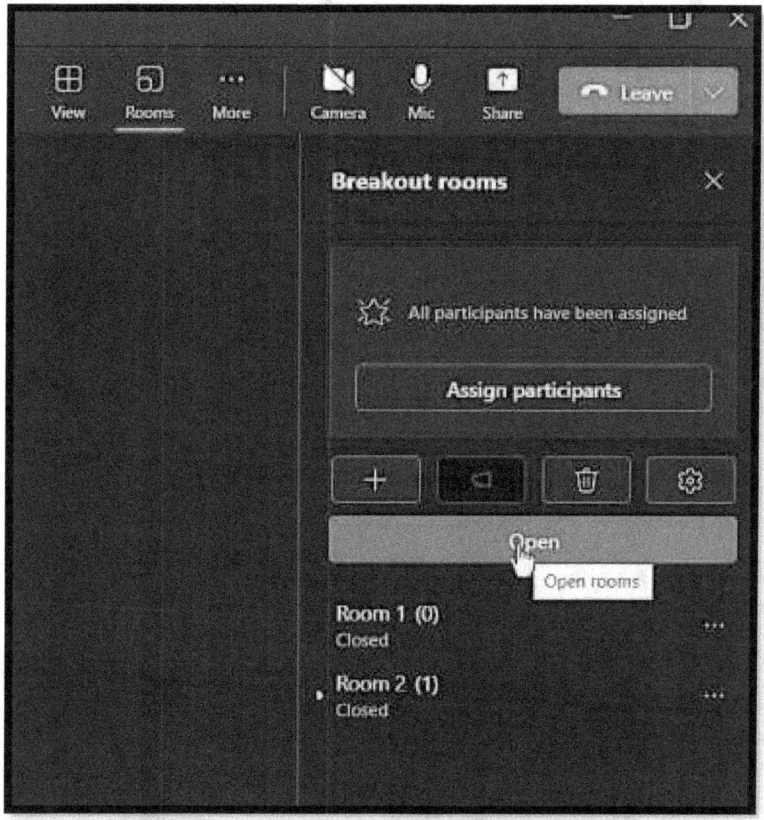

As the meeting organizer you can join whatever room you like. Let's say people are struggling with an exercise in room two, you could click on the three dots, say "Join room' and that's going to put you into the room as well so you can see it's opened up another window and then you can help out with whatever's going on there. You can also rename rooms.

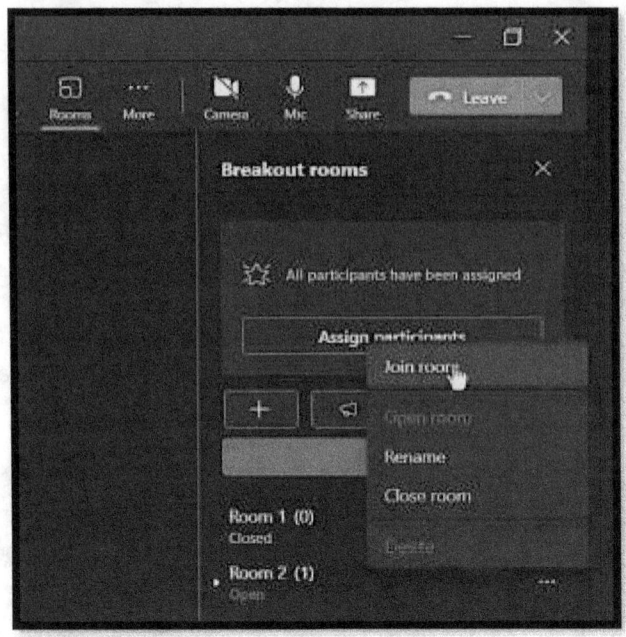

If as the meeting organizer you want to make a general announcement to everybody in all teams, this is where the "Make an announcement" button comes in. If you click on this, type in something and send that through that's going to be posted to each room, and then once it's time to close the breakout rooms and bring everybody back to the main meeting, you can simply click on the close button to close them off and you should find that in a moment all the participants are back in the main meeting.

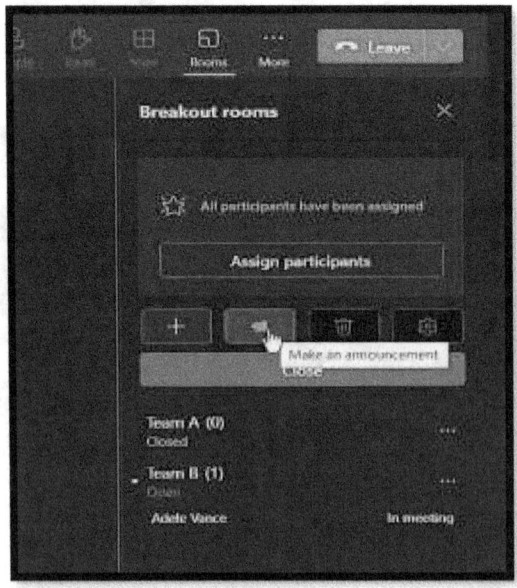

That's the idea behind Breakout Rooms. This works better if you have more participants but the key takeaway here is that you need to use the desktop version of Microsoft Teams to even see this Breakout Rooms option.

Review Questions

1. Open the Teams calendar and practice scheduling a new meeting.
2. Customize the meeting details, such as the title, date, time, and attendees.
3. Observe the meeting invitation and how it appears in the calendars of the invited participants.

CHAPTER 6
USING MICROSOFT TEAMS FOR PROJECT MANAGEMENT

In this chapter, we're going to show you how you can use Microsoft Teams for project management. If you didn't know already, Microsoft Teams has become the central hub for managing projects, programs, and entire portfolios. It brings together communication, content, tasks, meetings, and more to boost collaboration, productivity, and delivery. So in this chapter, we're going to be showing you how you can leverage Teams to master project management from end to end.

Structuring

Starting with structuring teams, channels, and groups, the foundation of Microsoft Teams is establishing your organizational structure correctly using teams, channels, and groups and this includes creating a program team. Start by creating a team at the overall program level. For example, if you're running an IT infrastructure program then you'll create a team with a similar name and this represents the overarching portfolio container for all of your related projects and the resources. The second thing is to add project channels. Within your program team, create dedicated channels to represent each project so based on our example; you may want to create channels for each project that rolls up into the broader program team. Think of channels as focused work streams under the whole umbrella kind of team.

The third step is building project groups. You can further organize resources within channels using groups and these allow you to bring together only the people needed for that project channel based on role, department, Etc. For example, for the channels we already created we might have additional groups and this keeps teams separated so the right people get access to the right channels and content. In summary, for the example project of IT infrastructure, that would be the program itself. For the channels, it would be individual projects, and for the groups, they would be Resources by role. This three-tier hierarchy keeps everything aligned yet separated as your portfolio scales up.

Communication

After that, comes the project communication using posts. The post tab within your team channel is the Central Communications hub for your project. Posts work like chat messages visible in real-time to your entire project team. Unlike email, posts enable quick conversations, questions, feedback, and status updates on the fly. To write a post, simply click the pencil icon at the bottom of the Post tab and this opens a text box. Here you can format text, @ mention colleagues, attach all kinds of files; add photos, and more using the formatting toolbar. When your message is ready click the arrow icon to post it into the channel and for the critical announcement that you want to emphasize, click the three dots on your post, select "Mark as important" and this flags the post with an exclamation icon to signal its urgency so people don't miss it.

After that, you can utilize getting feedback and approvals by @ mentioning stakeholders in your post. To get their attention right away and feedback for official approvals, click the three dots and select Ask for approval, Assign approvers, and Track status and if your channel gets busy then use the search bar to find posts by keyword or name. You can also save key posts as pinned for quick access later on. In summary, use posts for conversations, announcements, approvals, questions, status updates, and more. It's so much faster and more effective than email.

Sharing files

You can also use the Files tab within your channel to store and share all important project documents and collateral and this would include things like presentations, specifications, contracts, budgets and so much more. With Teams' built-in SharePoint integration, your files are accessible 24/7 to your team from any device or location with no more emailing files around. Set up a clear folder structure with subfolders to categorize and organize related project documents.

We would encourage using descriptive names and date or version numbers so the latest file is obvious as this would prevent confusion about which version is accurate. With shared files you get true real-time co-authoring, multiple people can work on the same file at the same time. Beyond just your team, you can still share files by clicking on "Share" and you will get a unique link to send to stakeholders outside your team for access but remember to set permissions, expiration dates, and password protection.

Integration of external tools

You can integrate external tools and applications. This is a major benefit of teams which is its extensibility, using tabs to integrate external tools right into channels. Some popular project management integrations would include Trello and Asana. You can also embed web applications, custom-built apps, SharePoint lists and so much more. This allows Microsoft teams to centralize work happening across a variety of platforms.

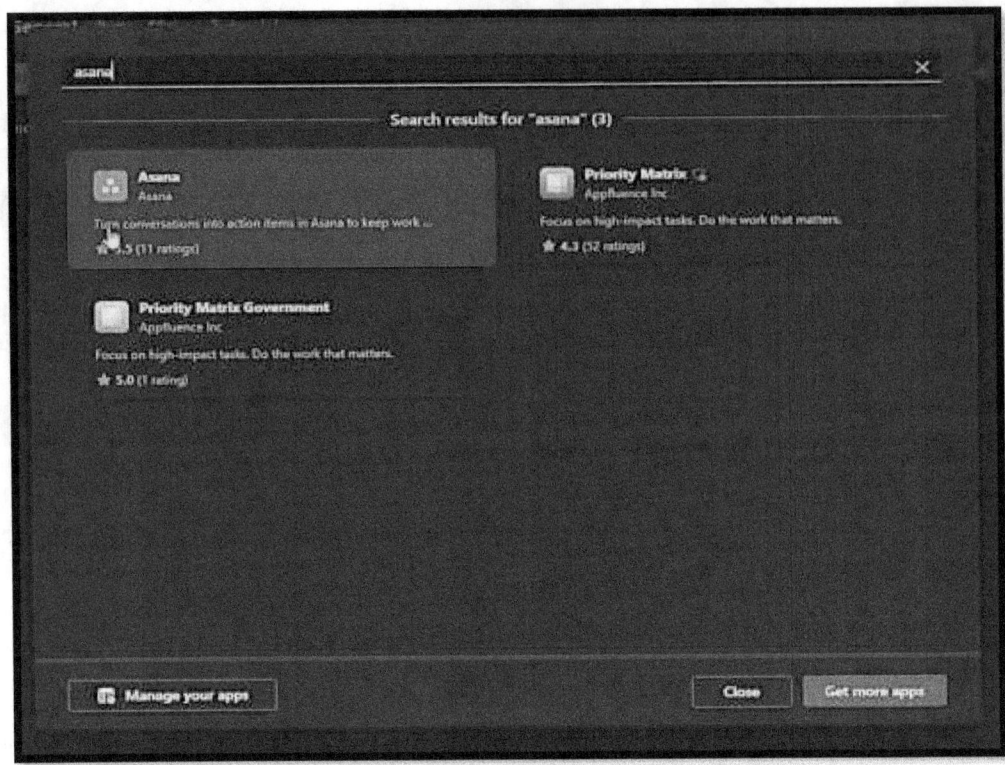

As you can see, using a free tool such as MS Teams could be a huge lifesaver when it comes to project management. To get started, establish your team structure, build out a sample project channel, and document the process.

Review Questions

1. Create a new team in Microsoft Teams specifically for a project management initiative.
2. Organize the team's channels and tabs to align with the project's lifecycle and communication needs.
3. Invite the necessary team members and assign appropriate roles and permissions.

CHAPTER 7
USING MICROSOFT PLANNER IN TEAMS

The new Planner available on Microsoft Teams harmonizes task management across all apps, bringing you a simple and easy-to-use task management solution available at your fingertips. It has a clean and simple navigation menu, combining Planner and To-do into one centralized area with three main sections.

My Day

Starting with My Day, this is identical to My Day in Microsoft To-do. By default, you will see all of the tasks that are assigned to you and done today and the source here indicates where these tasks originated from so if you hover over a personal plan it just says the name of the plan and that is a new personal plan feature that we will cover later in this chapter. Next, is a team's icon, and then after that we have a task from a task list within a Loop page. If you're wondering about how to manage the tasks that are assigned to you within Loop, we'll look at that.

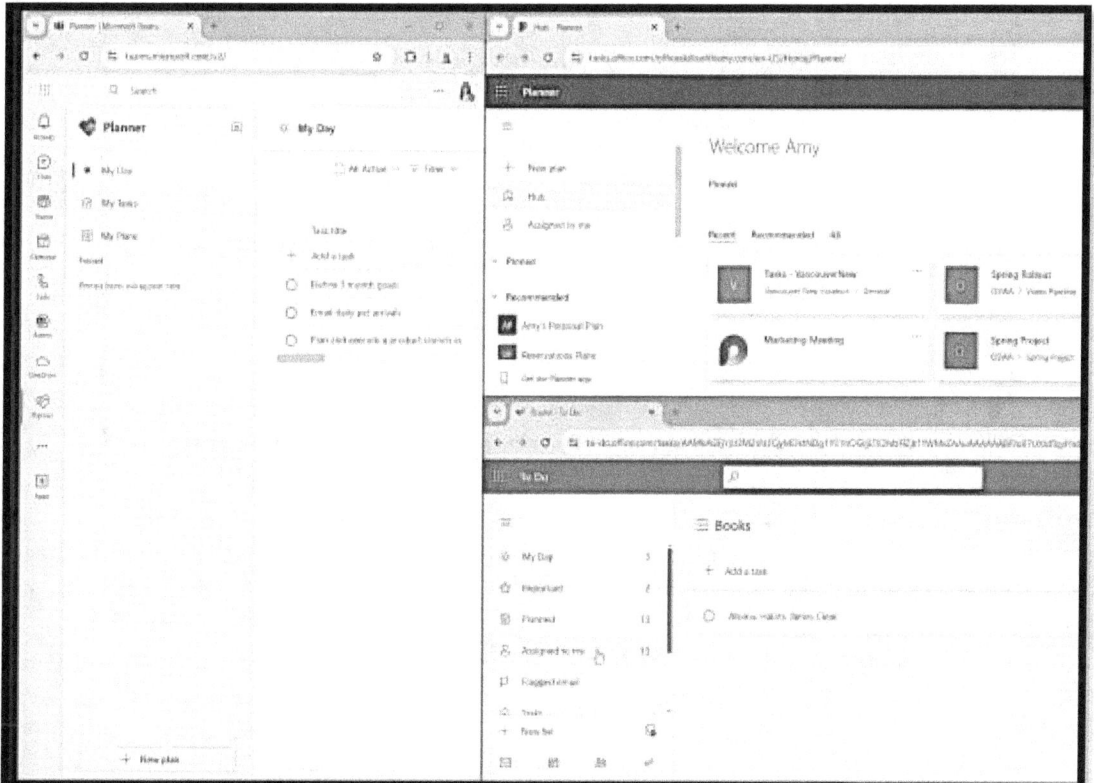

Using the illustration above, let's take a look at the first task there. You can see that it's got a little pink Icon as well as a notepad and sunshine. The sunshine relates to it being identified as My Day which is all of the tasks that are going to show up here but if we click into this task then you can see that it has a custom pink tag to it. These will be any color tags that are defined within any of your plans will show up there as well as the note here and that little notepad relates to the note which just indicates that if you open up this task you're going to find some notes.

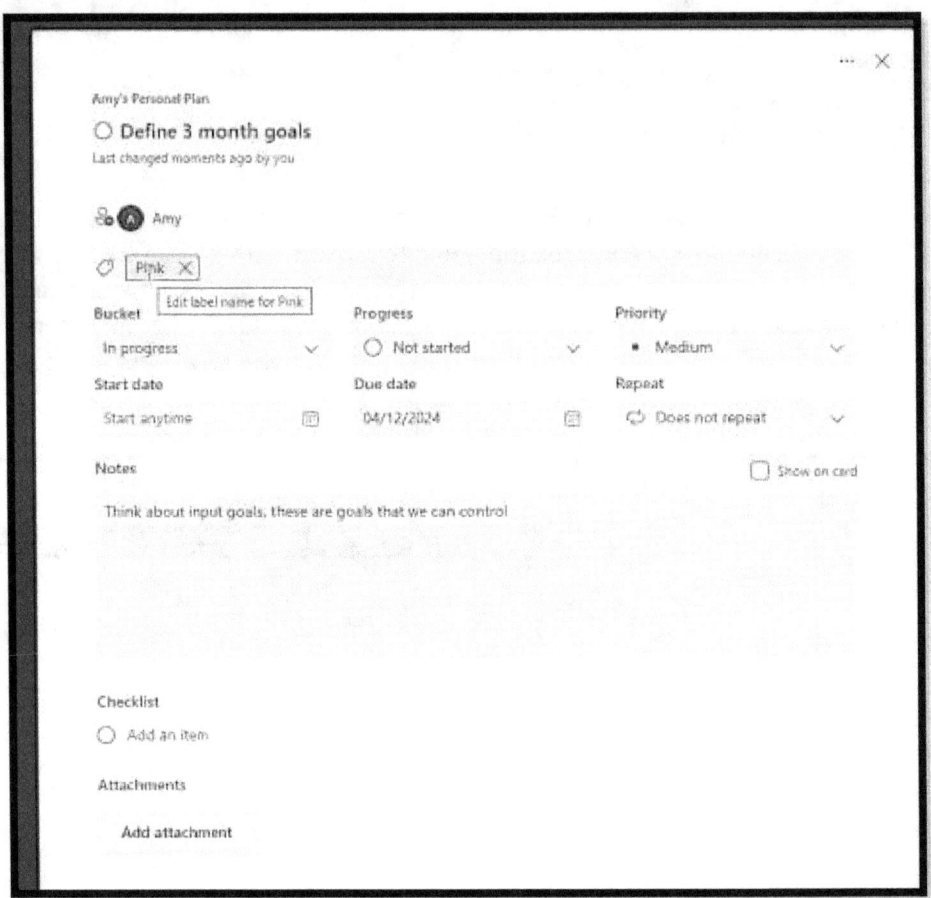

We can just create a progress bar and say "Hey this is urgent; we want to complete it today." If we close out of this then we can see that the little circle has been added which is the progress bar as well as the bell for the priorities so we can see how these create a nice visual effect when you just have that day view of all of your tasks. On the right-hand side, we have these ellipses and we can update all of those items here as well as from the top menu. They are identical, and wanted to draw your attention to the "Remove from My Day" button. Now this is super helpful if you have a decision paralysis so if that happens and you need to clear out some of those lower priority items you can simply just click "Remove from My Day" and now it is clear and you can focus on the tasks that matter most.

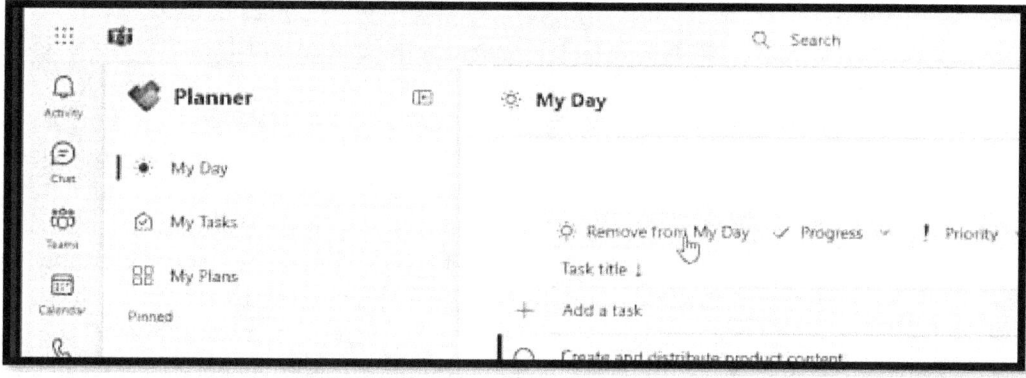

The last thing that we want to show you in this section is the ability to add a private task. You will see that a private task has been added and the source is a private task. These tasks are designed for you to quickly jot down to-do items at the speed of thought. To-do has a little completed section down here but we can easily replicate this by filtering for completed tasks from the top drop-down menu.

My Tasks

Moving along to My Tasks. This is a centralized area for you to view all of the tasks that are relevant to you and there are four predefined filters at the top some of them may look familiar from Planner and To-do.

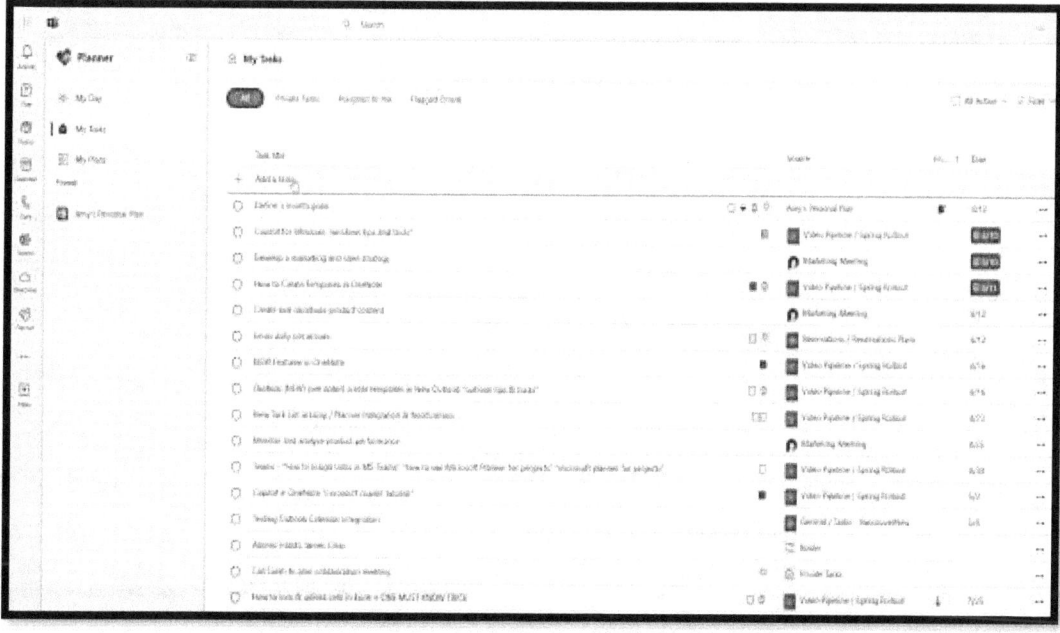

Starting with the Private task, we can see that private task that we added under My Day and that little Sun icon is there as well. We are also able to add new tasks here and if we want to assign that to today then we can just select it, click "Add to My Day" and it will show up under My Day. "Assign to me" is a centralized area for you to view all of the tasks that are assigned to you across all of your plans for all of the dates. Now depending on the number of tasks that you have you can easily sort by title, priority, or due date and there are also some further filter options for dates and priorities on the top. Moving over to Flagged emails, here we are in Outlook and we can just click this "flag this message" option and here we have that flagged email. The source is showing us Outlook and we've even got an attachment there. If we click and open up this task we can just head on back to that email by selecting the attachment here. My Tasks is all of the tasks in one place. Even that flagged email is showing up here and if we complete this email we will see it update in that little tag area with an Outlook as well.

My Plans

Moving along to My Plans, this is a centralized area for you to access all of the plans that are relevant to you. This includes lists from to-do plans within Planner as well as the new personal plans feature that we're going to cover in just a moment and soon to come will be premium plans which are going to be a new subscription.

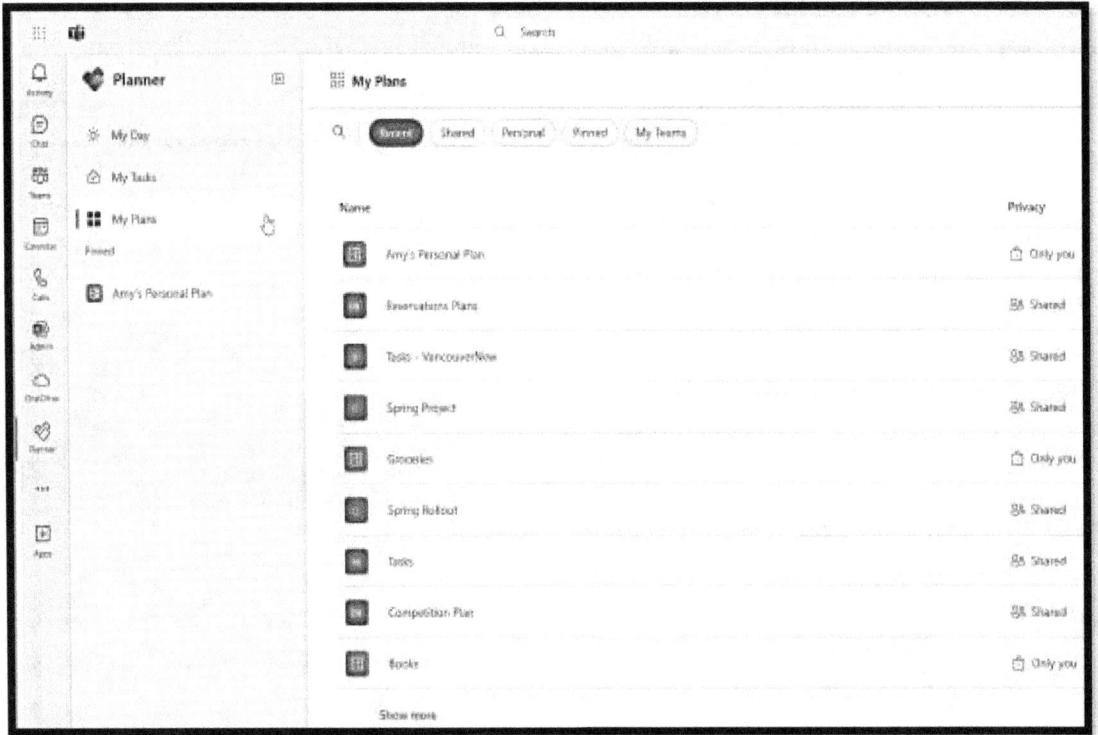

Starting with Shared plans, these are plans that are shared with you and are associated with a 365 group including a Microsoft team. Select "New plan" and head on down to see all templates on the left-hand side. Here, we have all of these different templates that include a preview of the plan, and these all indicate basic here but in the future, you might see some that will say premium and those are going to be part of a different licensing.

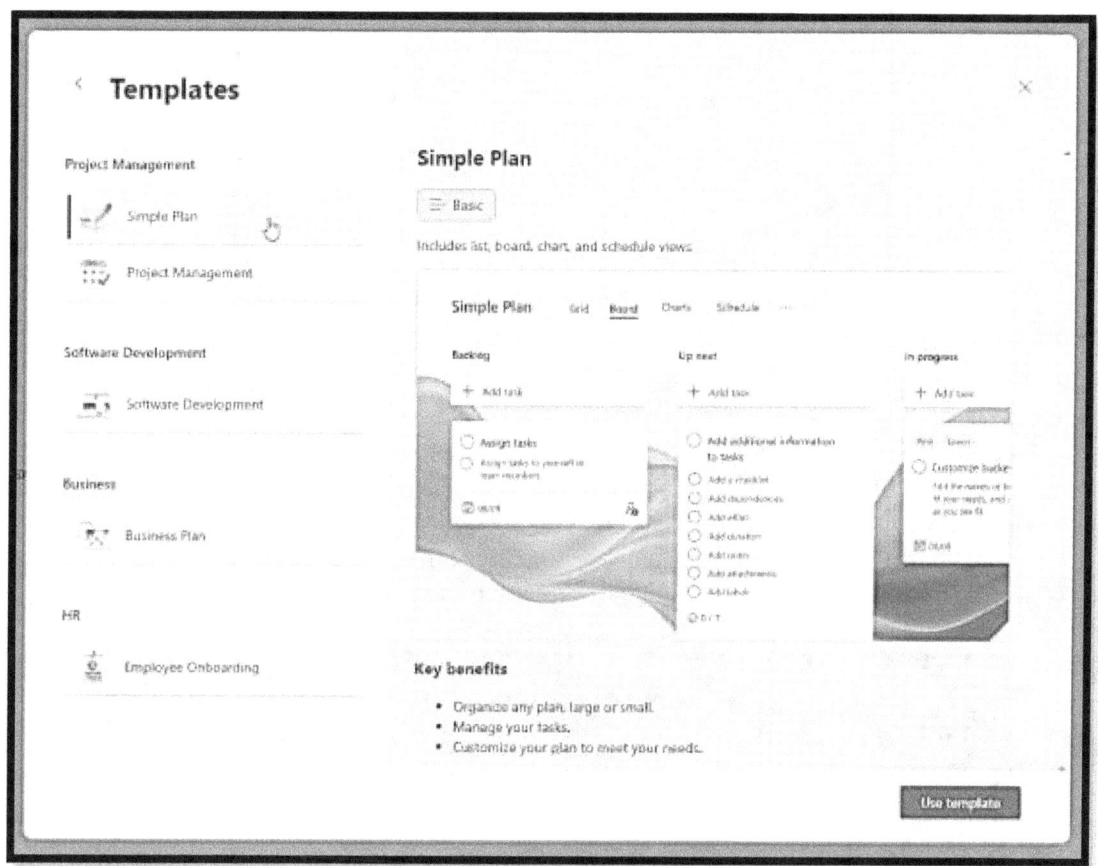

Let's say we are starting a new pet food line, we'll go ahead and select project management, for example, and click "Use template at Amy's animal shop." This plan is going to be for managing that product project. Since we're going to be using this plan quite frequently we'll pin it to the navigation menu.

From the bottom here we are going to select a group. You'll see here that a lot of these are your teams within Microsoft Teams and there is also an indication of it being public; some of yours might say private so these just indicate the privacy settings for that group. Let's go ahead and assign this to the operations and here we have our plan that has just been created.

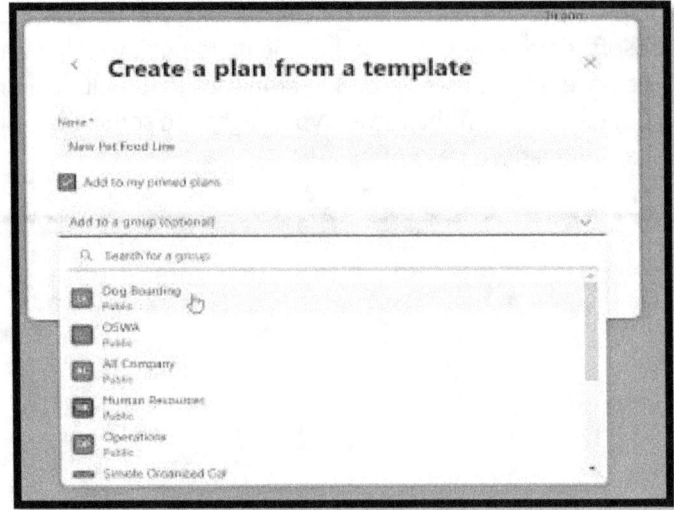

You can see that this amazing template provides such a good foundation for starting this project. You can also see that this plan has been pinned to the left navigation menu and if we head on up to My Plans then we can see that this new pet food line plan is showing up under the shared area. Moving along to the Personal tab this includes your personal plans, that new feature that has just been released with the new Planner, as well as Lists from Microsoft To-do. To create a personal plan we are also going to head up to that New Plan button and we have all of those templates available but if we want to start with a plan from scratch then we can just select Basic here.

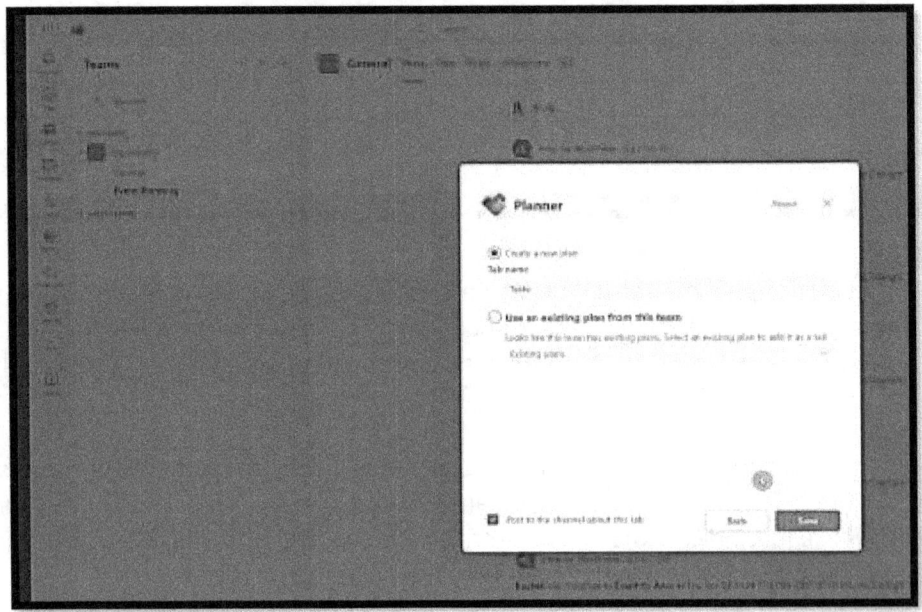

At the bottom where we assigned it to the group in the previous step, we're going to leave this blank and that means that this plan is not going to be unassigned to a Microsoft group. Let's go ahead and click Create and we now have our new plan. This is just so amazing because it never used to be possible. We used to always have one of these plans assigned to a group but now you can have a private plan so you can create your own plan and enjoy all of the benefits of a regular plan but having it be just for you. Later down the line if you decide that you want to add this plan to a group then you can create a new group or even add it to an existing group then once again select from the drop-down menu and add a team. Our Pinned Plans are the same as the plans that are on the left navigation menu. You can pin or unpin them.

My Teams

Moving on to My Teams, here we have all of our teams that have plans associated with them and if we expand them then we can see the plans broken down by channels within those teams. We want to do an overlay here because the Shared Plans and My Teams look very similar, however, you're going to notice that that new pet food line as well as this other plan that we just created are not showing up under the My Teams. If we look on the right-hand side here under the "Shared With" we can see the team that they are assigned to but not the channel whereas My Teams has these plans that are broken down by Channel.

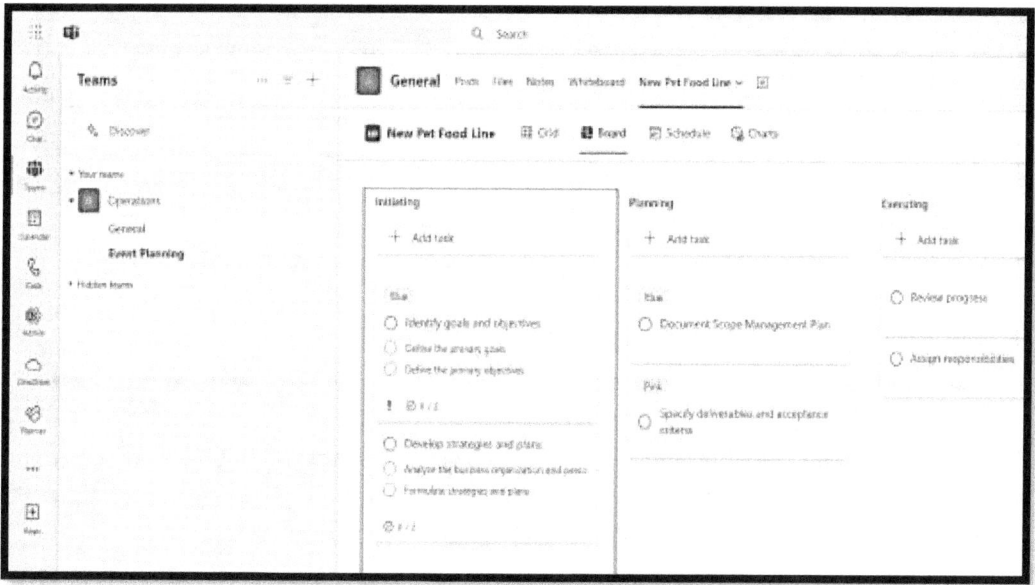

What we need to do is add these plans into a channel for them to show up under the "My Teams" section. So here we are within that team and if we head up to the "Add a Tab" area then we can see the Planner right here (if you don't see it then you can just search it right here). Now we just need to add our plan. We could create a new plan or we can use an existing plan from this team so we've already assigned that plan to the team, we just need to add it to a channel and there we now have that new plan that we just created. We'll go ahead and click Save and we can see that the new plan has populated here. If we go back into My Plans and My Teams then that team is now showing and we can see that new pet food line at the bottom.

Important settings

There are some settings that we want to highlight. From a plan, if we select the plan name at the top, we can rename our plan from the right side here, and at the bottom is the ability to delete a plan from the group area. This is where you can change the settings of the 365 group and then we have notifications as well.

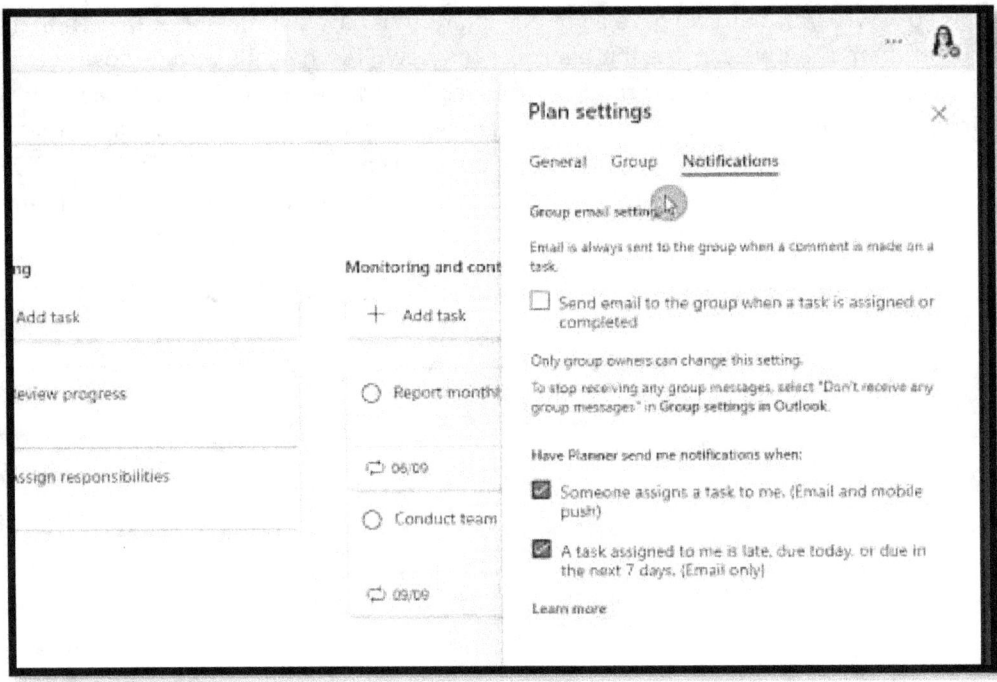

Just note that for this first one here only owners can change that setting but here you can define if you want to receive emails and mobile post notifications when someone assigns a task to you or if a task is late due today or due in the next 7 days.

Old vs. new planner

If we navigate to the old Planner then there are some settings here that we would like to see in the new Planner. From the ellipses on the top here, there are a couple of pretty cool things. The first one is exporting the plan to Excel which creates a nice little table, next, we can share the progress of this plan with somebody who's not directly involved with the details, and also the ability to add a plan to an outlet calendar which is a way of integrating a plan directly into your calendar that you might look at a little bit more frequently.

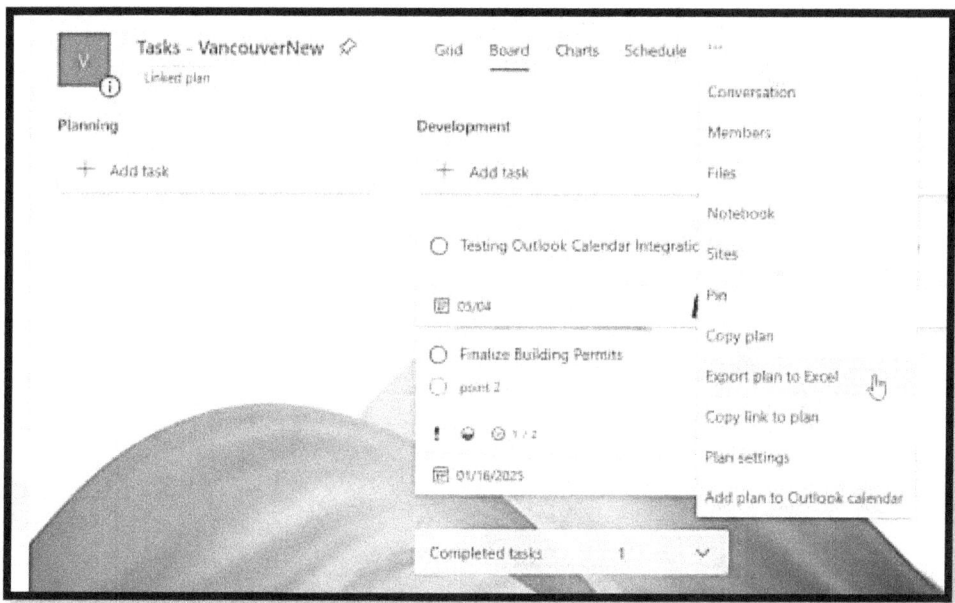

Now if we head into Plan Settings, similar to the new version we've got that General group and notifications tabs at the top but under the General we have these beautiful backgrounds which are not yet available but hopefully will be soon within the new Planner.

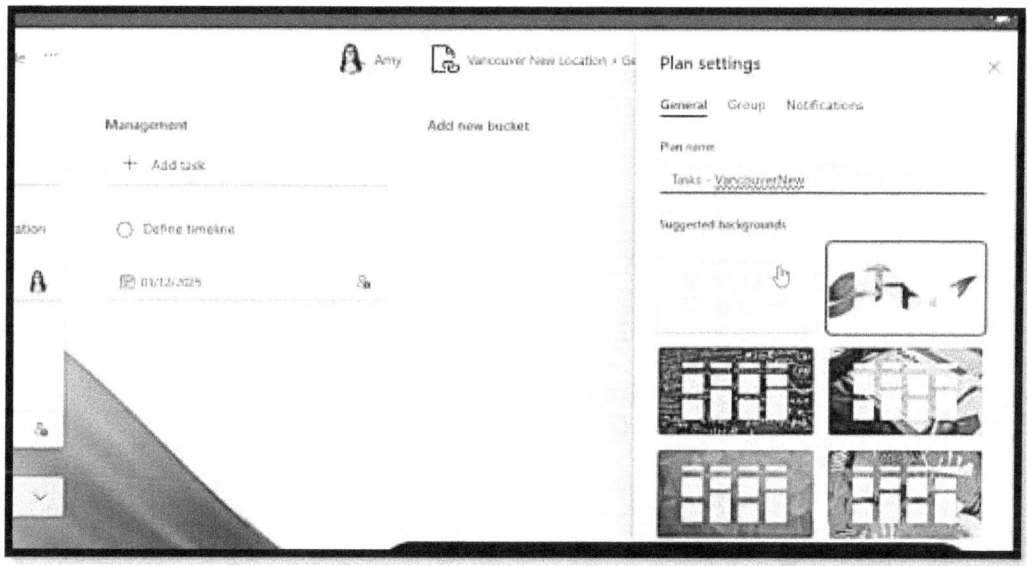

Review Questions

1. Navigate to the "My Plans" section in Microsoft Planner.
2. Customize the plan views to suit your preferences.
3. Observe how the "My Plans" feature provides an overview of your involvement in different project or team-based plans.

CHAPTER 8
MICROSOFT TEAMS FOR EDUCATION: EXPLORING NEW FEATURES

In this chapter, we'll be showing the new features in Teams for Education. This includes annotating PDFs for educators and students, time-saving assignment updates, Reflect improvements, and a whole lot more.

Annotate a PDF

The first new feature is a longtime request from Educators and that is the ability to annotate a PDF in an assignment. You'll go to the Assignments channel as an educator, click "Create" and then "New assignment." Give it a quick title and instructions. For example, let's say TPS report Explorations and we want the students to add their own TPS report to the assignment and to make it a PDF. After that, you're going to click "Assign."

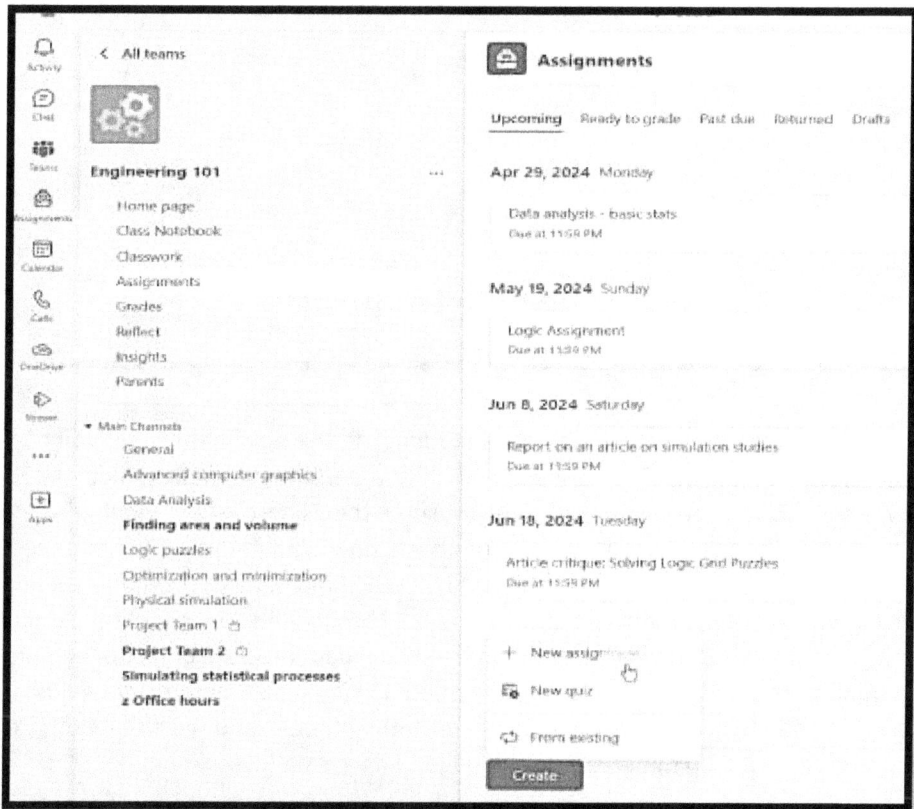

Now we'll flip over to the student. We're signed in as a student and here we'll find the assignment. We're going to open this up, click "Attach" and we're going to add the PDF to this assignment.

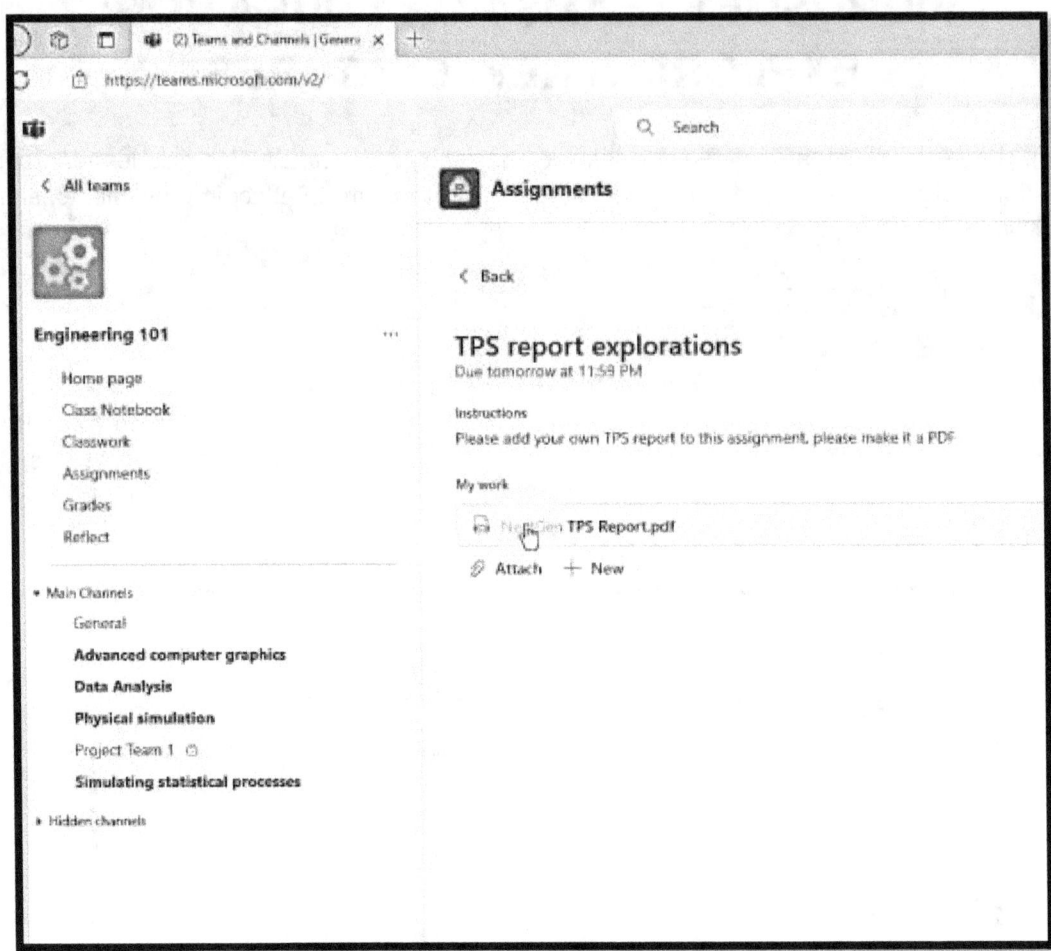

We can open this up and then click "Edit." As a student, on the right-hand side, you can see that you can edit and annotate your PDF as well. We'll click a pen here, choose a nice color, and sign our name with the mouse. You can go and highlight some things here as a student or you can erase them so you have quite a few options as a student and now we will click "Save changes" to save these back to the PDF then we'll click "Close" and our PDF is all ready to turn in to the teacher. To do that, we'll go to the upper right and click "Turn in." Now, we'll switch back to the educator to show what it looks like in the speed grader. So we're signed back in as the educator and we're going to open up the report then we'll scroll down and open up that assignment that was just turned in. We can go again and click "Edit" just like we did with the student and we'll pull out our nice red pen and now we can mark up right here on the page.

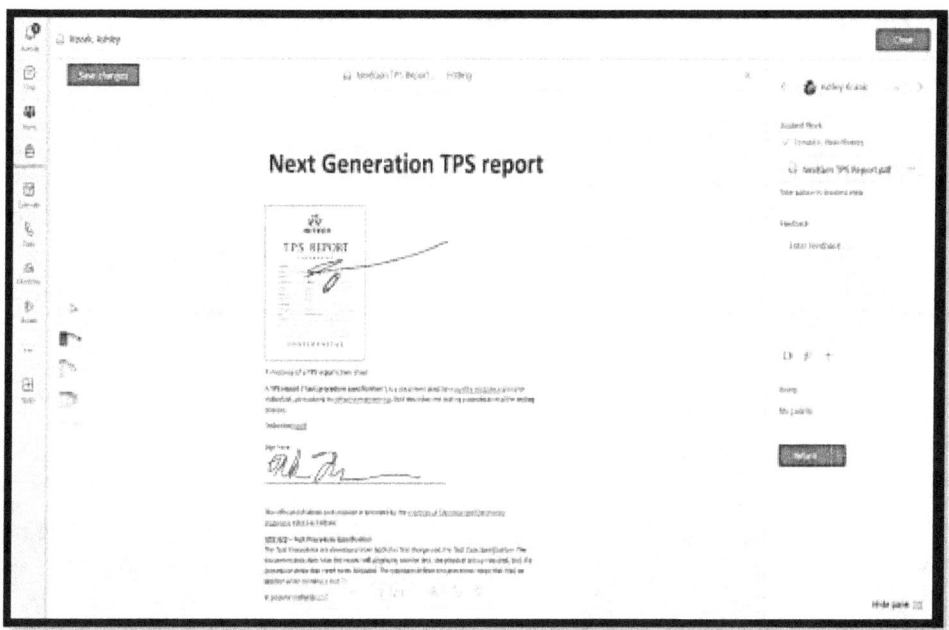

We can point to the arrow right there on the report, scroll down, and circle some things and you can do this with a device but using a mouse you can do all the same things that we were showing before now. Now as the educator though you can do this in your grade and when you're done, just go to "Save changes" and move on to the next student.

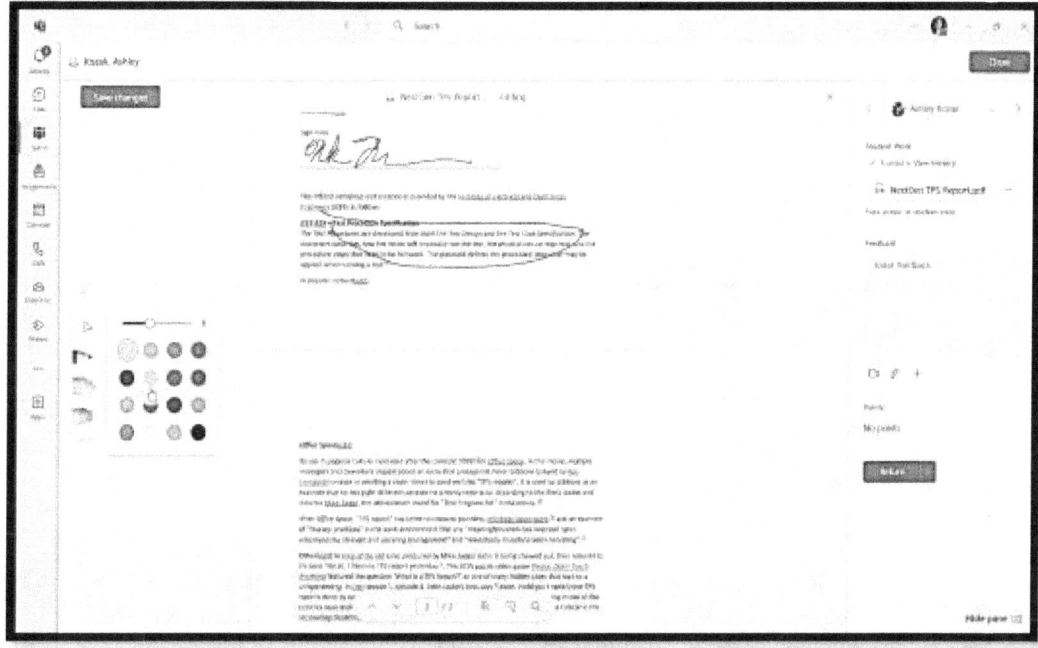

Reminders

The second new feature allows Educators to easily send little reminders to students who might have late assignments. You'll open up an assignment here that you've already sent out and here you'll see a set of students that you still need to return it to if they've not turned it in. Select these students and when you select them you're going to see a couple of options. One of the options is to send a reminder.

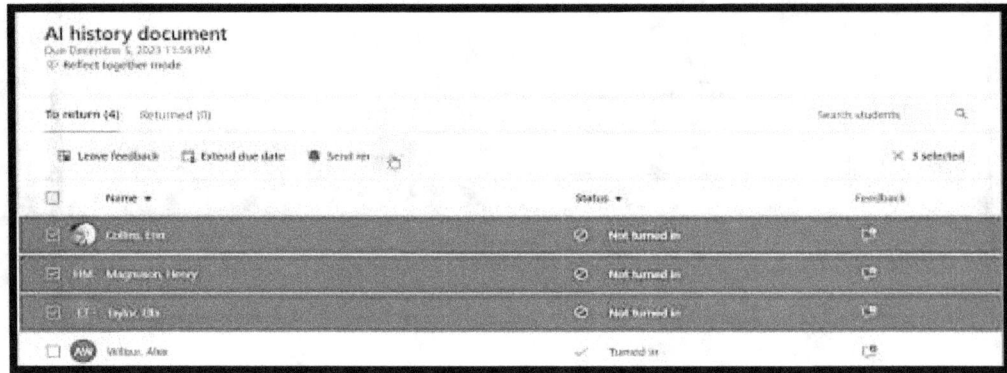

Click that and it'll give you a confirmation - a notification will be sent to each of these selected students reminding them and this will show up in their Activities. Click Send and now all those students will get a ping in Teams.

Extend due date

The third new feature is in the same area as the second one. For this illustration, we're going to select a student here and we want to give her an extended due date. When you select that student in the list here, you're going to get "Extend due date" and because we want to give her a little more time we'll select a different due date, enter the date and time, and then click "Done."

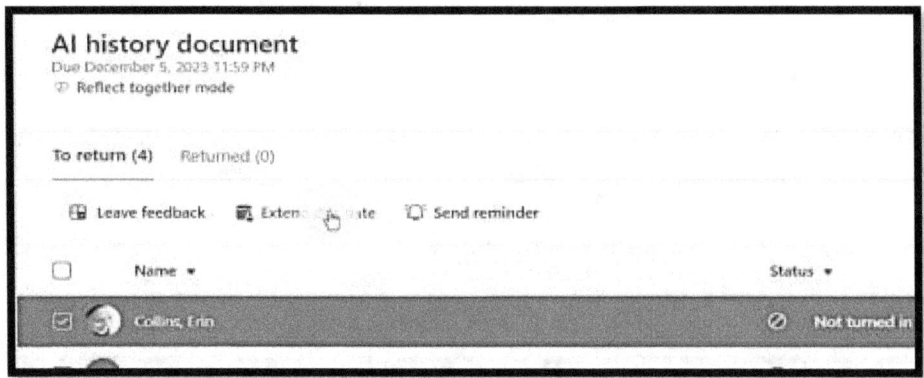

Now because this student has a special due date it shows up right here so you can know that the student has an extended due date, whereas the rest of the class has the original due date.

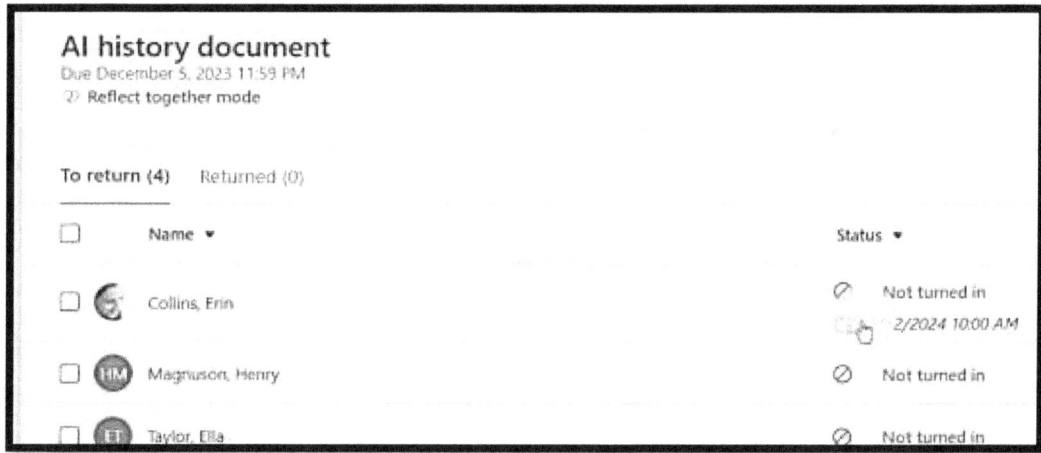

Bulk feedback

The fourth new feature is also in this assignment list. You can select a couple of students and leave bulk feedback at the same time. Let's say you want to leave the same type of feedback across for many students you can just go to this option, type in your feedback and click Done.

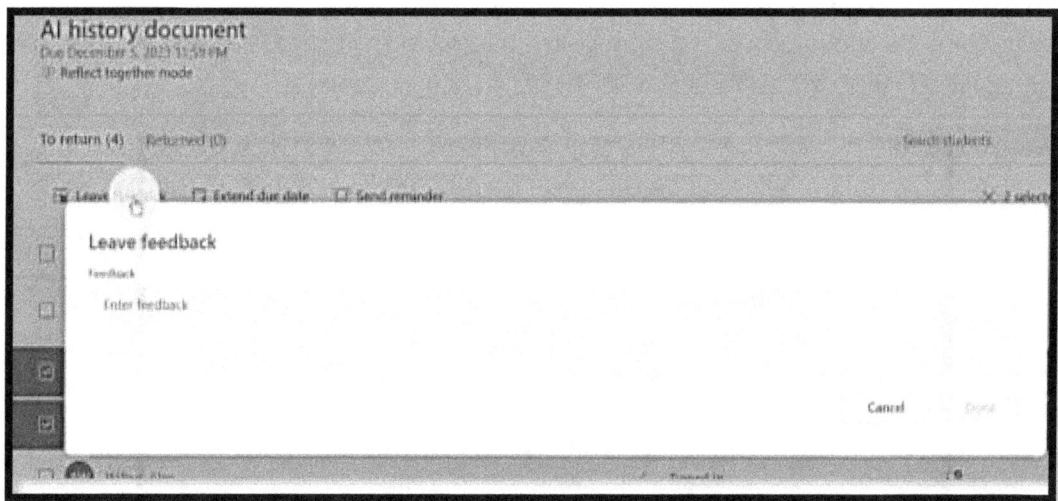

What happens is that it will put it across the board. You can use this feature in the case where the students didn't turn in an assignment or if they've all turned it in and you just want to bulk leave something across the board.

Reflect mindful coloring book

The fifth new feature is the Reflect Mindful coloring book. For this illustration, we're signed in as a student and we will go over and click on "Reflect." This brings up the students' version of the Reflect feature and here, you'll find a new option which is "coloring book."

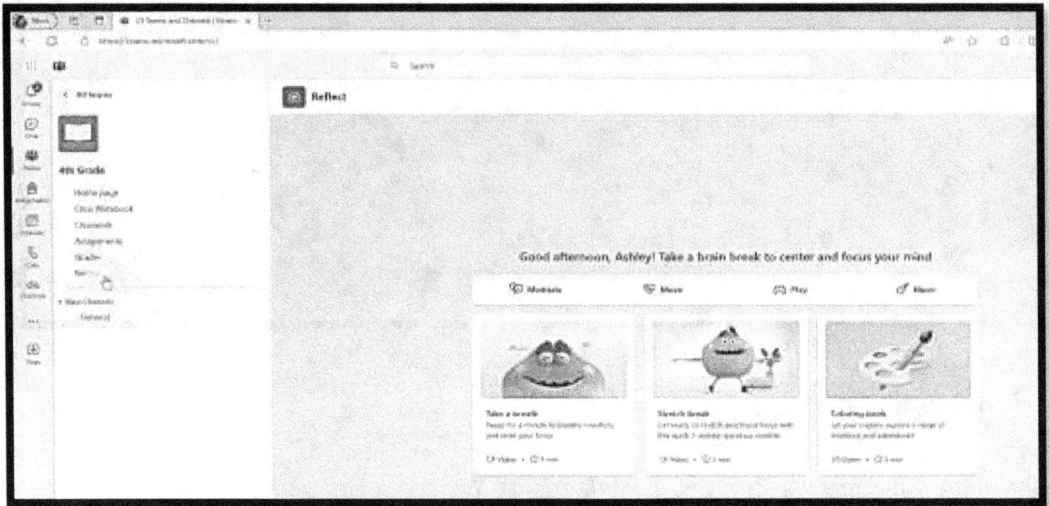

This will take you to the redesigned student homepage with lots of fun things but the newest one is the coloring book so we will click on that and there we'll find different emotions along the bottom including the ones about focused, frustrated, and happy. You can choose a page that you want, and there are lots of different options - you can go random or choose one.

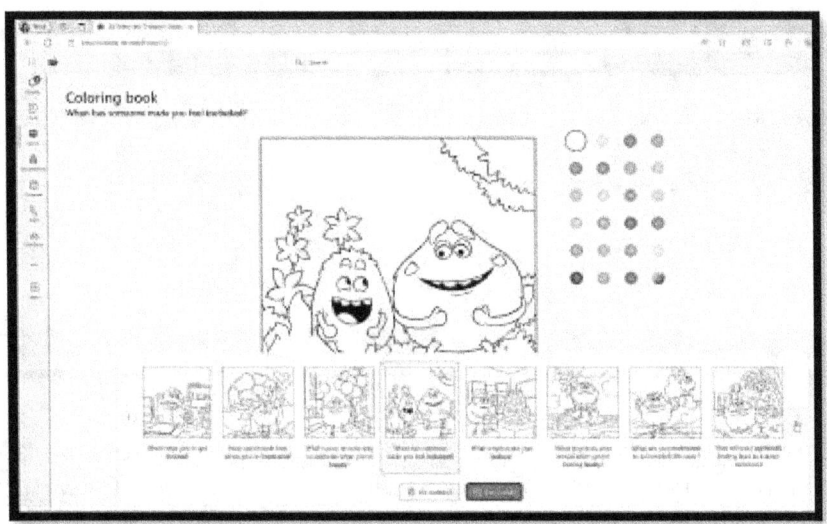

We're going to do "Why did you feel successful recently" then we'll go over this emotion and start clicking on the color. We can play with colors by giving him a red tongue and a nice sky-blue background with the white clouds and we can click different parts of it so it's very easy to color different aspects. It's very quick and easy to start making a fun design and we're going to finish this up, get a few more little places and what's nice is when you're all done you can say "I'm done," Start over or Change the picture.

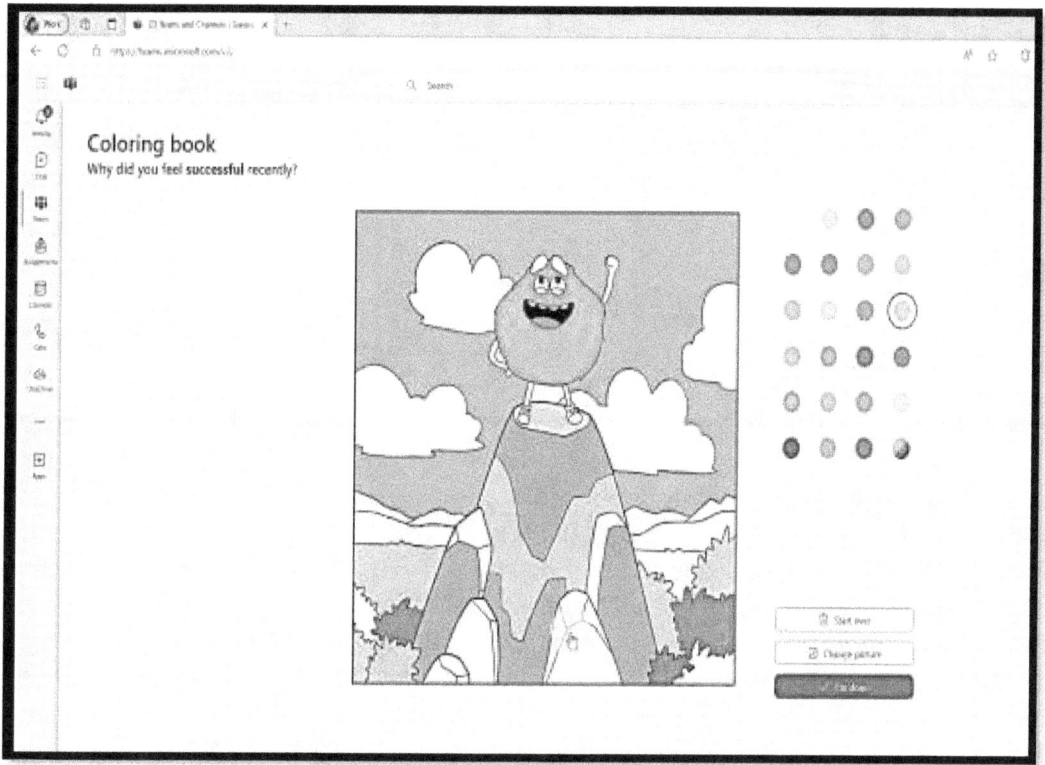

We'll click "I'm done" and then it gives us a little celebration. We can say "Save this" and it will offer to save this as a JPEG that we can open up. We can share it or choose to try another picture. You can see that this mindful coloring book is a fun and relaxing way for students and adults to have a lot of fun, using the Reflect feature.

Reflect for staff

The sixth new feature is also in Reflect but it is Staff teams getting Reflect and that is Reflect for grown-ups. For this illustration, we'll go into our staff team, and when you create a new staff team, as the owner, you're going to see Reflect at the top.

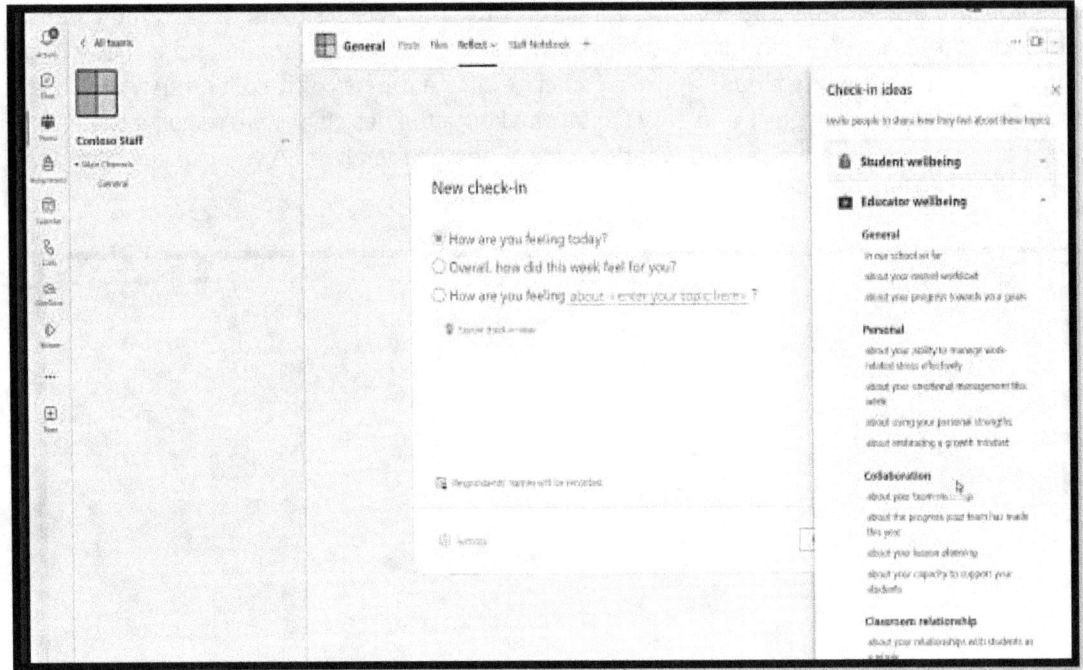

As the staff account owner, you get to choose the Reflect check-ins. Click into here and just like you've seen in the class, there is a Staff Reflect. Click on "New check-in" and the options to explore check-in ideas come up. When you open that not only do you have Student well-being, you have Educator well-being and these are specifically tailored for staff. You can ask about how things are going in your school so far, about your workload, your ability to manage work-related stress, team meetings, and the progress your team has made this year. If you click an option it adds to a check-in and you can do a preview. So just like you've seen in the regular Reflect which is a similar type of Reflect, this is where staff can choose these same types of feelings monsters then you can hit "Submit." All of these things operate in the same way that you would normally see with the Reflect in the classroom so with the Reflect feature for staff you can create this check-in and capture all this information for your staff.

Noise suppression for reading progress

The seventh new feature is noise suppression for reading progress. In Teams, reading progress is one of our learning accelerators that help with reading fluency. As the educator, go into assignments and open up an assignment from a student. Let's say there is a reading assignment your student turned in that she recorded herself reading out loud but there is a background noise and it throws off the auto-detect AI that helps things like mispronunciations, insertions, and self-corrections.

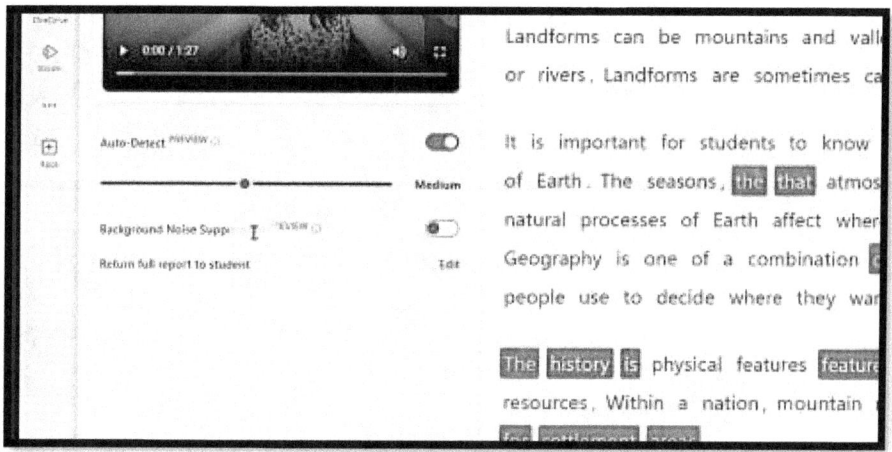

What this feature does is add background noise suppression so what an educator can do is if they want to have an AI layer applied on top that does background noise suppression it can give a cleaner reading of the student. If you are an educator and you want to turn this on, just flip this Background Noise Suppression button to ON and what it'll do is recalculate the entire passage and remark it based on the noise suppression layer.

Warning for missing attachment

The next new feature is a warning for students in cases where they forgot to turn in their attached work. This is a great timesaver for teachers so they don't have to nag the students who forgot. As an educator, click "Create new assignment," and give it a title and instructions. What you'll do next is use the words attach or add or attachment as keywords in your instructions so when the student is turning it in, if they didn't attach anything you will give them a little prompt.

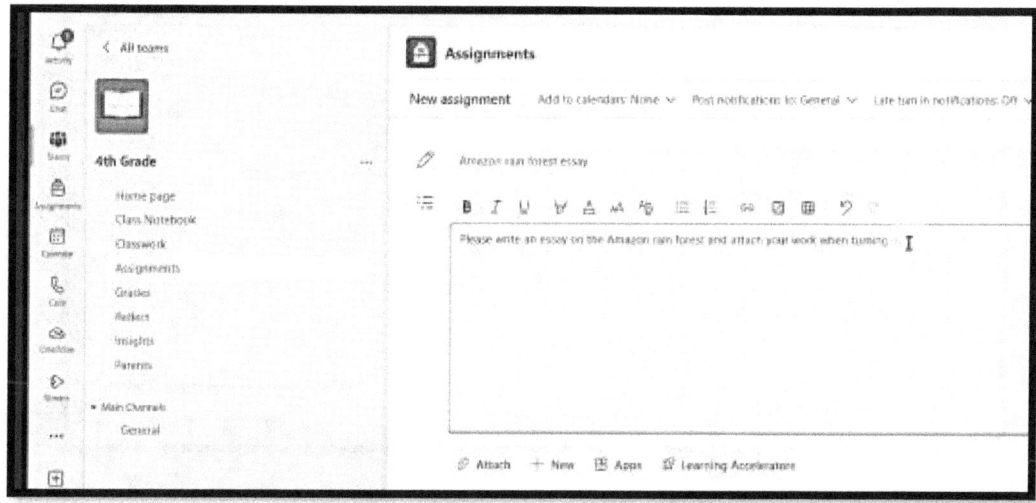

How that works is after sending out the assignment, let's say the student opened up the assignment and forgot to attach any work, then turns it in, they'll get a little error message notifying them that there's no work attached. That student can now click the Cancel button, go back and attach it, click Turn in and it's all set.

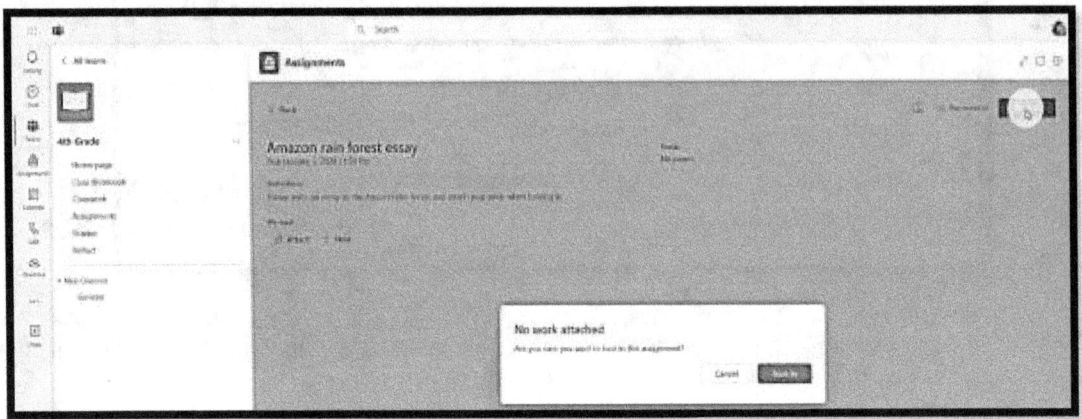

Updated editing option for students

The ninth new feature is an updated entry point for assignments to make it easy for students to edit their own work. Let's say you're signed in as the teacher, click "Create," "New assignment," then click New, choose Word document, give it a title, and click Done. In the past, the only way you could make it so students could edit their own copy was to hit the three-dot menu and say students can edit their own copy but then some Educators couldn't find that very easily. What they've done now is expose it right on this page so it's right in your face; you can go here and say students edit their own copy and now when you send an assignment, students will get their own copy to edit.

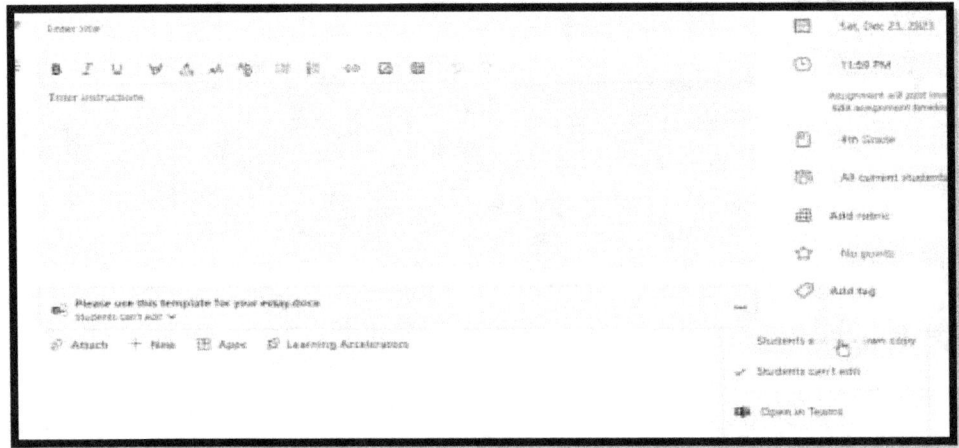

104

Use of tables

The next new feature supports tables inside of assignment instructions. After giving a title, when you click there's a little table button. You can click this and say insert table. You can decide to add more columns if you want to make it bigger, you can insert a row, format the table, delete the entire table, or delete columns and rows.

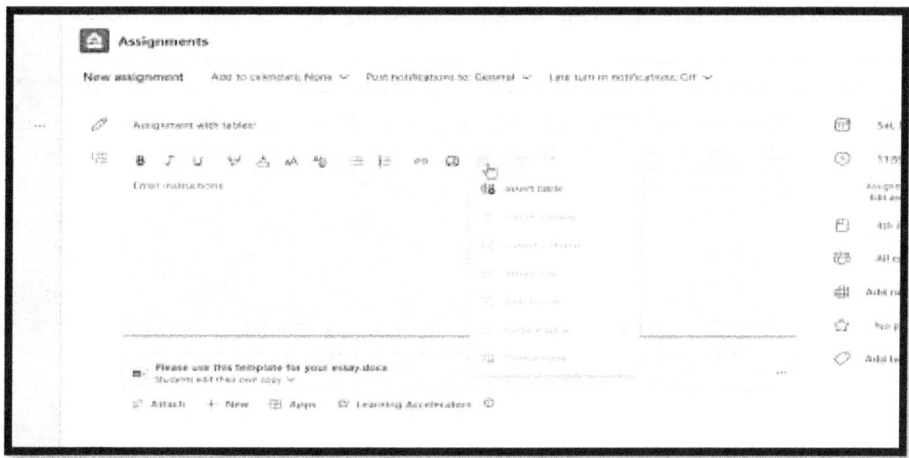

After inserting your table, you can go ahead and fill out your table putting whatever you want into that table. If you want, you can also size it to make it bigger.

Updated Turn in celebrations

The next new feature is updated Turn in celebrations. If you're signed in as a student, as you open up an assignment and turn that in you can see an example of one of the newest Turn in celebrations and it's different every time for the student. In the upper-right click "Turn in" and you'll see there's a new little celebration.

School connection app

The last new feature is the school connection app which is built into the Teams' mobile consumer app on iPhone or Android. It lets Educators keep in touch with parents on how students are doing. Note that this has to be enabled by your admin and what you'll see is a set of students and their map to different parents and Guardians. The way to do this automatically is this: if your school uses School data sync and you have those parent emails in the school data sync, that will map automatically so everything will be set up with the students and those parent emails.

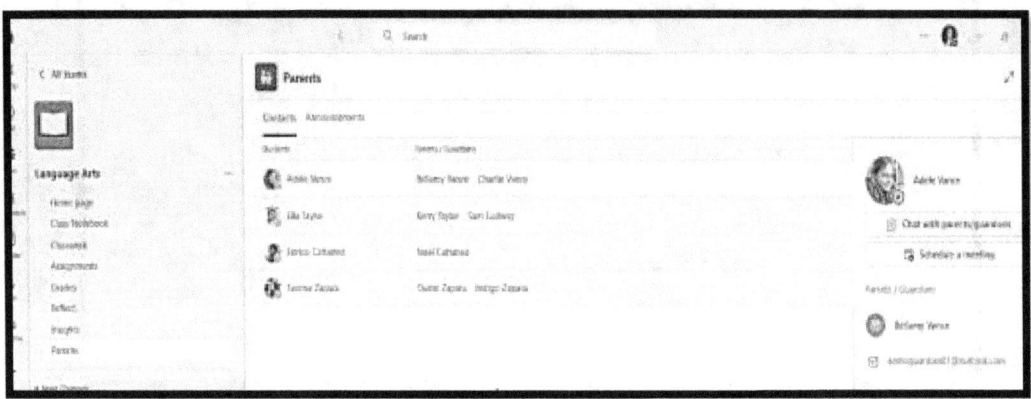

Now for this illustration let's say a parent has already been mapped to a student. You're going to switch over to the mobile phone as a parent and launch Teams for Education on your iPhone. Next, you're going to choose your guardian account and it'll sign you in.

Next, add the school connection app; this is the consumer version of Teams so in the upper left side tap on the little profile that's the name of the parent and goes into the settings. Now tap settings and then at the bottom, you'll see the school connection option.

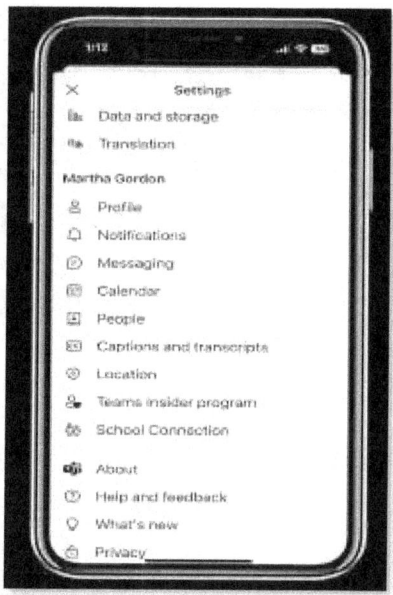

Tap that option then go to "Enable," turn this on then hit back and close that dialogue. In the middle of the lower part, you're going to see a little school connection icon, and little parent and the student icons. Tap that then load up the school connection - this is going to pull in the information for your child.

You can see how many assignment activities your child has, you can tap on that child and it's going to show more information about what's happening in the class. All this information is available to you as a parent.

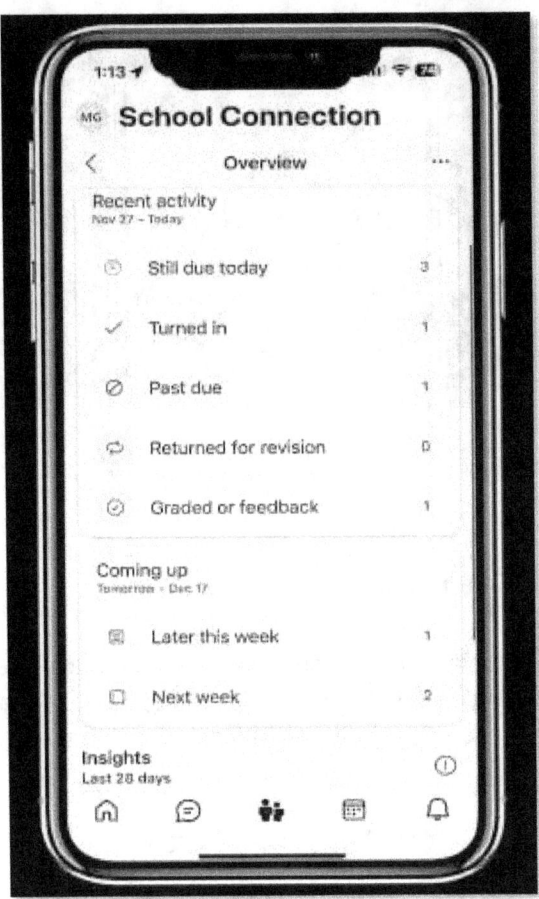

You also have Insights bottom that enables you to see what assignments are turned in. If you have learning accelerators like reading progress you'll be able to drill in and see reading accuracy, new practice words, and what type of digital activity your student is doing.

Review Questions

1. In a Teams channel or chat, upload a PDF document and practice using the annotation tools to highlight, comment, or draw on the document.
2. Observe how the annotations are displayed and shared with other team members.
3. Discuss the benefits of the annotation feature for collaborative document review and feedback in an educational setting.

CHAPTER 9
TIPS AND TRICKS

In this chapter, we're going to cover some Microsoft Teams tips and tricks along with some basics. We'll also go through some of the new apps appearing in Microsoft for 2024.

The basics

The basics of Teams are: set up a large team larger than you think you need, try and get away with one team only split out where there's privacy needed and the rest will be easier or irrelevant because you'll be working more out in the open. Speed up time to decision-making and there's one place to do everything. There's one place to do everything or one place for your files. Channels is the first place that you will start your file structure because every channel sets up a folder in the SharePoint site that gets set up when you set the team up. After you've done that then the main Basics that make everything else easy in Teams is to only use the teams bit of Teams and don't use chat. If you've got something one-to-one that's private or not work-related, chat is fine for that but a lot of organizations get tied up in knots by overusing chat so if you can move into Teams' channels that will make your life so much easier and when you're in a channel, assuming that you are used to either using chat or using teams with not a lot of people in there's a couple of things that you might need to change your behavior to get the most out of using Teams channels. Microsoft Teams used to have a little text box down the bottom and it was quite difficult to see whether you were replying to an existing thread or starting a new post. That's now gone because there's a "Start a post" button and if you click "Start a post" it now pops up with a lot more options than it used to do. We used to have to click a little button to make this box bigger to add a subject and type a message, in the new Teams; they have helped us out with that so we don't need to do that anymore.

Now if you're starting a new thread always add a subject (which as we said earlier was a bit more tricky to get to) so that when you're in a thread of information you can quite quickly see what that thread of information is about. Always add an @ mention into a new post or no one's going to see it. If you want to put something into a channel Just for information and whether people see it or not it's fine then don't put an @ mention but if you're in a normal organization, you're doing a new post and you don't put an @ mention, no one is going to see your post so you always want to do that. If you just start typing it's going to suggest some things for you. If you want to do a channel mention you can either start typing @ Channel and it'll mention the channel you're in or you can do @ and start typing the name of the channel. Everybody who hasn't hidden that channel will get pinged about your post. Now assuming you got your teams and channels set up correctly, then people would have shown channels that they work in and hidden channels that they don't work in so anything hidden channel is for private information. You can do individual @ mentions as well so always start a new thread with a heading and an @ mention. If there's an existing thread then click reply. Those are the basic things you need to know when moving to Teams channels away from chat. Chat pings everybody in the chat by default, because of that most people mute a lot of chats and then it's difficult to get things through so at mentioning again with a reply is going to Ping everybody in that thread of information above you by default but they could have turned their notifications off for that thread so make sure that you're doing an @ mention in a reply as well if you're applying to a specific person just to make sure that they get it.

Mute unrelated conversation threads

If you do have a long thread of information, let's say that a thread was started, you added something in and then that conversation sort of moved away from what you put in or you just wanted to throw in your opinion into that thread, you can click the three dots and then if it wasn't your thread and there were other people in there, there'd be an option to mute this conversation thread. With that, you can turn off notifications just for that specific thread.

Create a task from a post

You can create a task from a message. You can also create a task in the planner which is right here in the More Options section and that will then link back into that conversation thread so you can see the context of what was going on and can write a task for yourself to do with that conversation thread in Teams. You can either put that into "at the time of recording" which is changing shortly as the planner changes. At the time of recording can either put that as a personal task which will go into tasks only. You can see that by default unless you choose to share it or you can put it into a planner board which in this instance is living inside a team so you can put it into your planner board and that's shared for everybody to see that's it within that team that's got the permissions to see it. Those are the basics that you need to get the best out of Teams channels.

The activity center

If you look at the image below we've got the activities pulled up.

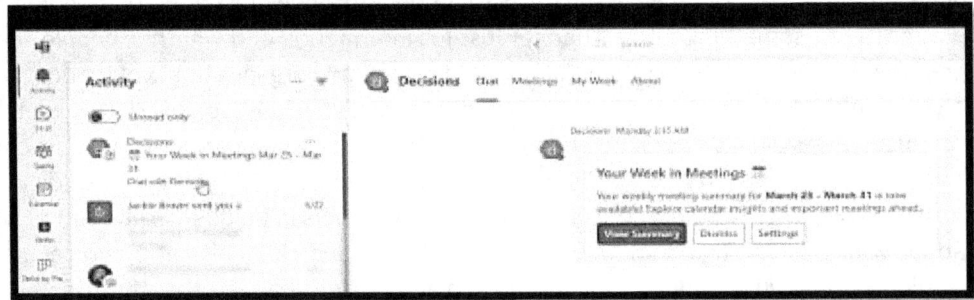

These activities are what have happened recently that maybe you haven't seen. If you click on this little button right here at the top left-hand corner, you can check all the unreads and then if you click it again it'll show everything, even the ones that you have read. Something cool about this is you can filter it so if you click on these three lines on the top right you can choose to type in some sort of filter. Let's say that you only wanted to see things that came from Bobby, you would type Bobby and notice that it would only show the activity for Bobby. This activity center shows the updates for the teams' channels for your chats and anything that goes on inside of any of those.

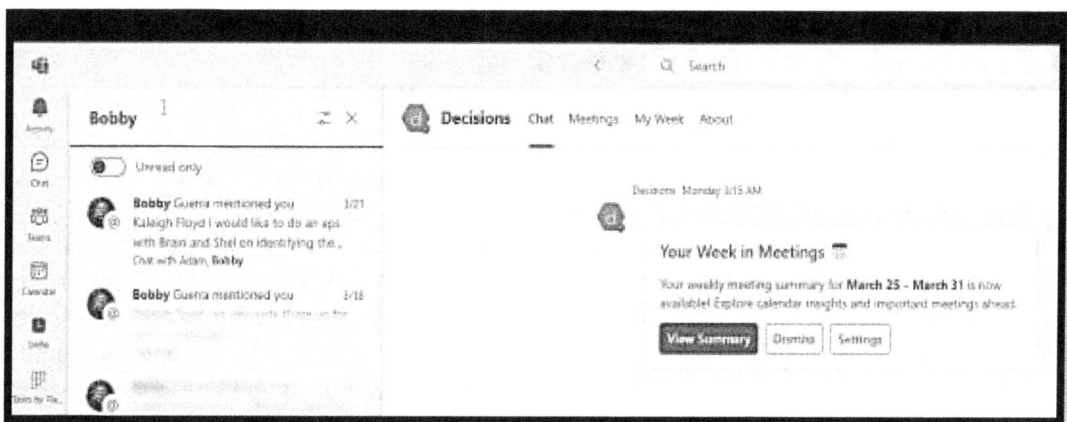

Another cool thing is if you look here on the "more filters" option it shows examples of filters that you could do; you could even filter it by an app. Now if you exit out of that and you go to the three dots next to activity you've got some more options here. You can go to the notification settings and inside the notification settings, you've got some options here to play sounds with the notifications, play sounds with incoming calls or meetings, and show messages and content previews on the notification.

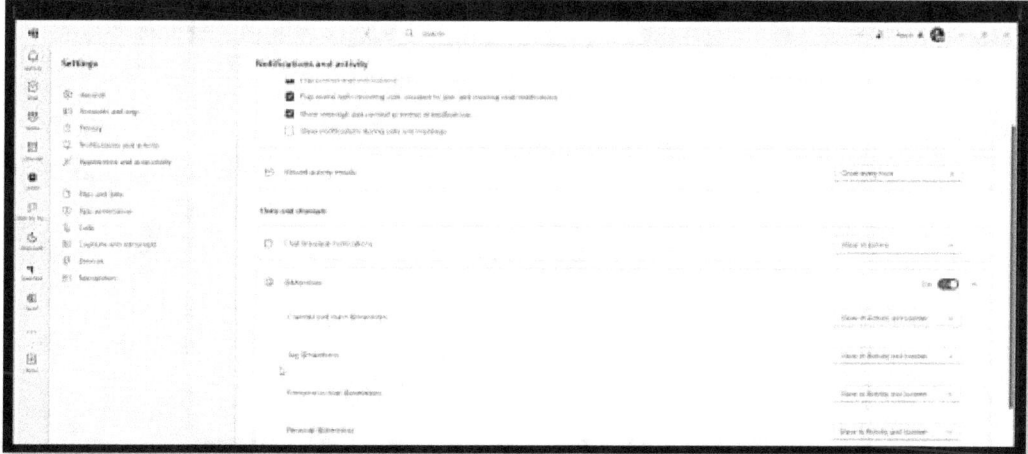

We would highly suggest if you're just starting to use Microsoft Teams go into the settings area and make sure that all of these notifications and activities are set to your preference. For example, for the chats and channels section here you can choose to show messages as a banner or you can choose to turn it off so chats won't show up on your screen. You can also choose if you want to be notified when somebody mentions you inside of a channel and a team tags you in a chat or anything like that. All of these are important to make sure that you get the correct notifications you want and that you don't get the ones that you don't want so make sure to go in here and check and make sure that yours is how you want it before you get started with using Teams.

How to set your pop-up windows

Now while we're here in the Settings area we want to go over to this General tab. This is important because they've done some pretty cool new things with the most recent Microsoft Teams update and these are the options to create Windows. Something cool is when starting a new chat or opening content inside Microsoft Teams you can choose to open a new window which means that it'll pop out a brand new window for that specific thing. Let's say you're inside of a chat with somebody and they send you a link to an Excel document. Before now, it would take you inside of Excel inside of Teams, it's like you're in Excel but you're also still inside of Teams and this could be quite annoying to some people. The new option is that they have pulled up Excel in a new window when you click on the link and to be able to set that up you would click this "new window icon when opening content inside of Teams" then if you want to set up a new window every time you start a new chat you can come down here and click New window here and every time you create new chats this will pop up a brand new window for only that chat and only that person that you're talking about. It could be a group chat as well.

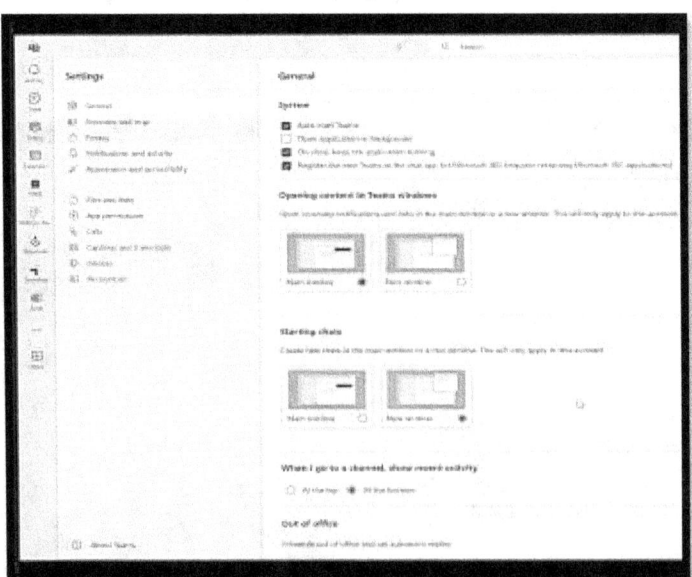

Schedule an out-of-office notification

The last thing that we want to mention inside of this general Settings is you can schedule an out-of-office and this is super helpful. If you're going to be out of town for a week and you want to let your team know and remind them about it you can choose to schedule that here just by clicking "Schedule," turning on automatic replies and you can set an automatic reply for the out of office message. You can also choose if you want to send replies outside of the organization or only keep them internal and you can also send replies only during a period.

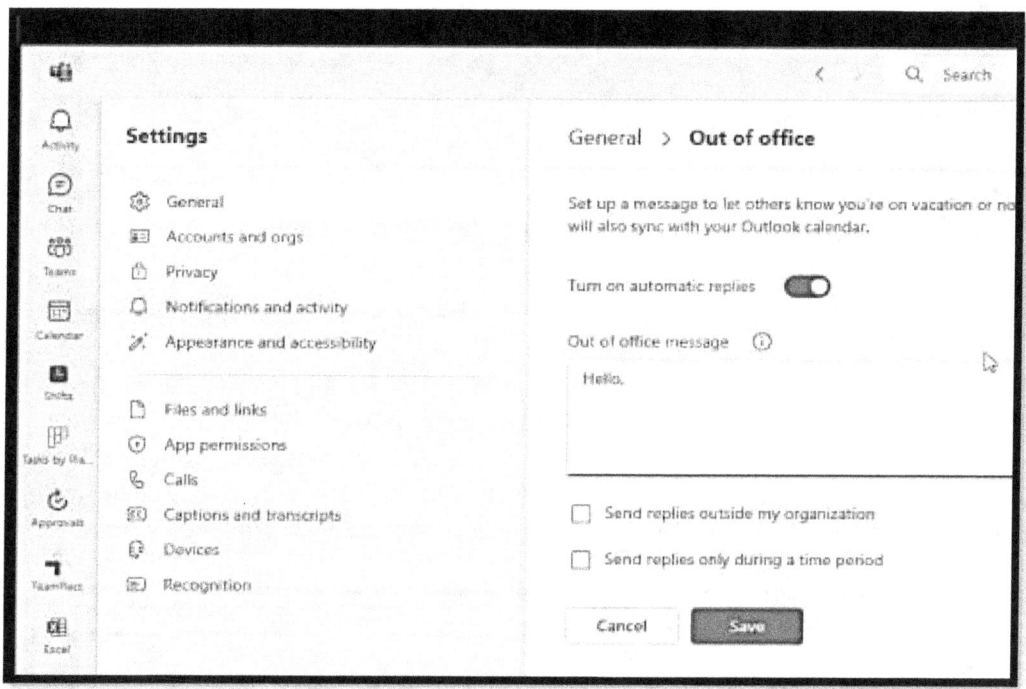

When you click on that you would set when you're going to be on vacation and when you'd want this out-of-office message to send to people. There are more settings that we can talk about but those are the basics that we would suggest taking a look at when getting started inside Microsoft Teams.

Adding new tabs

Let's talk about adding new tabs at the top. In the general Channel, you have three tabs (Posts, Files, and Notes) and then next to that, you have a little plus. Let's say you want to add a Whiteboard tab to this channel so that people can brainstorm ideas and scribble down notes, you can do that very easily by clicking on the plus icon.

This is going to show you a long list of all of the different apps and if you click on "See All" you should be able to find Whiteboard in there. Click to add it, give your whiteboard a name, or leave it on Whiteboard and post to the channel about this tab so that everybody knows you've created it, then let's click on Save. It's going to create the tab and it's going to load up a blank whiteboard that everybody can use as they would if they were working within the Whiteboard app itself.

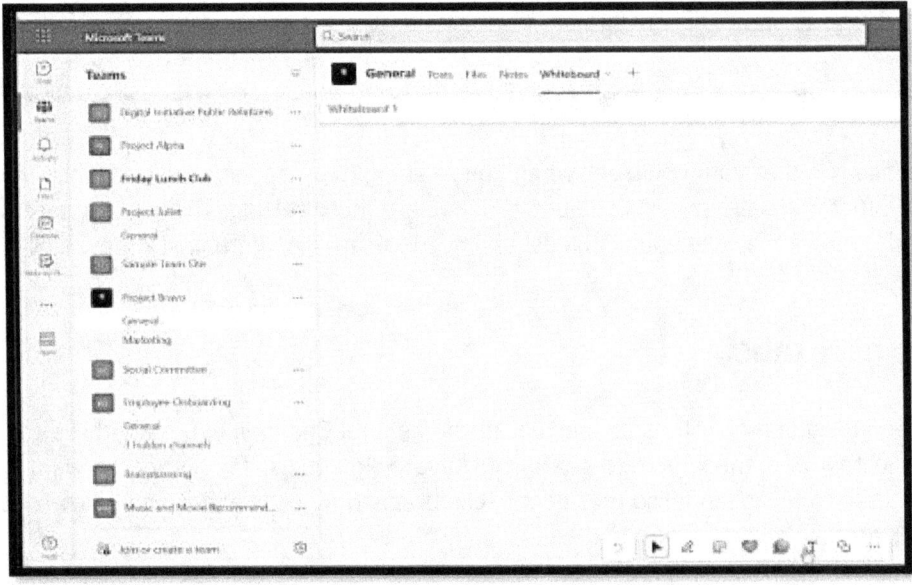

You'll notice you have a little floating toolbar at the bottom. With that, you can add things like notes, text, images, and comments. What about an actual file? If you click the plus icon again and let's say you want to permanently share an Excel file as a tab, click on the Excel app then go in and search for the Excel file that you want to use in the tab then click on Save. That's going to create a new tab with the name of that file and it's going to load up that Excel spreadsheet into that window. This is a nice way of keeping all of your files together within the channel so that everybody can easily see them and also collaborate on them as well adding apps as tabs is a great way to customize your channels.

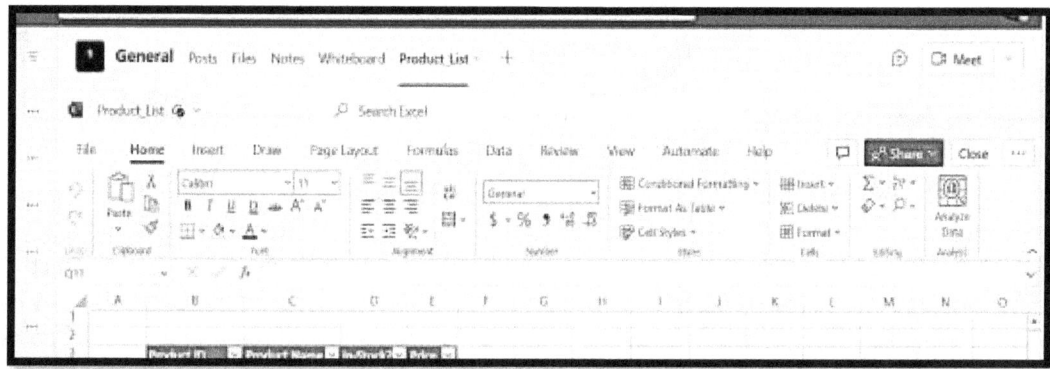

Using the search bar

The next thing to talk about here is the search bar at the top because there are a couple of little things here that are going to make your life a lot easier. You can use this search bar to search for anything within teams whether it's a file, a person, a team, or a channel.

115

If you click in the search box and look at the wording there it says to look for messages, files, and more or type forward slash for a list of commands. So if you type forward slash (/) this gives you access to a bunch of shortcut links to execute specific tasks. For example, if you want to quickly change your status from Available to Away, you could type **/away** or simply select it from this drop-down and when you hit Enter you'll notice your status has now been set to Away. If you do the slash again you can set it back to Available very quickly.

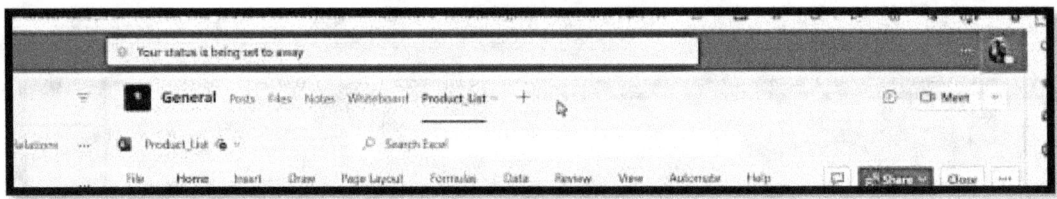

There are quite a few things you can do here and some of them will require your input. For example, if you want to send a quick message to a person you could type in **/chat,** and then it's going to ask you which person you want to send a message to. After selecting that person. You can effectively type the message into the search bar and send it through and because that message is just to that person it's going to show in your private chats. This forward slash is great. If you want to quickly jump to a specific team or channel you have a **/goto** option just here and then you can select your team or channel. One that we find particularly useful is **/saved** because that will give you a quick way to jump to all of those posts that you've bookmarked. Now aside from forward slash you can also use the @ symbol here as well to quickly do other things. For example, you can @ somebody to send them a message so this is very straightforward but you do have other things you can do with this @ symbol as well, one of which is that you could send somebody some praise. If you click @ praise, it's going to access the praise app.

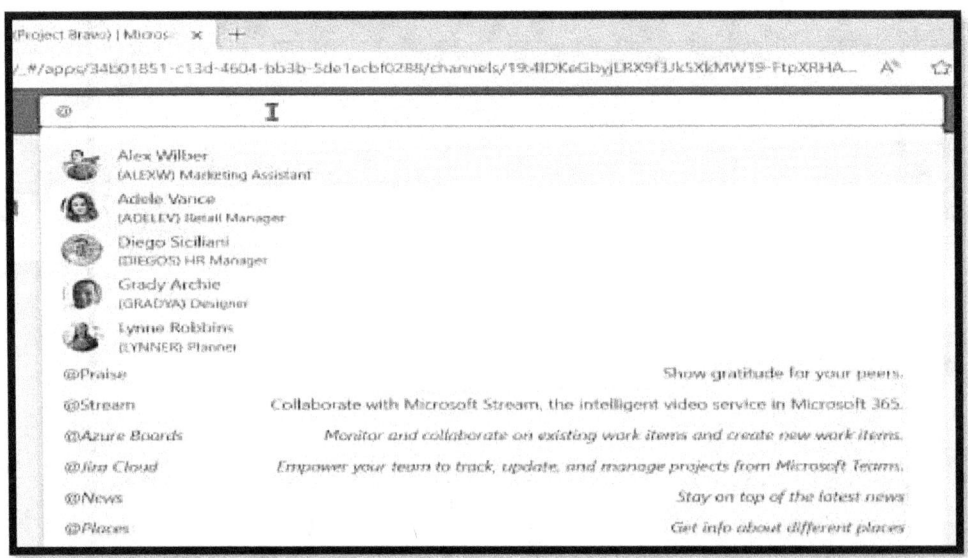

116

Let's say you want to tell Mary that she is awesome, you can add a personalized note, change the background color, and send that through to Mary. That's quite a nice way to boost someone's morale and let them know that you think they're doing a great job.

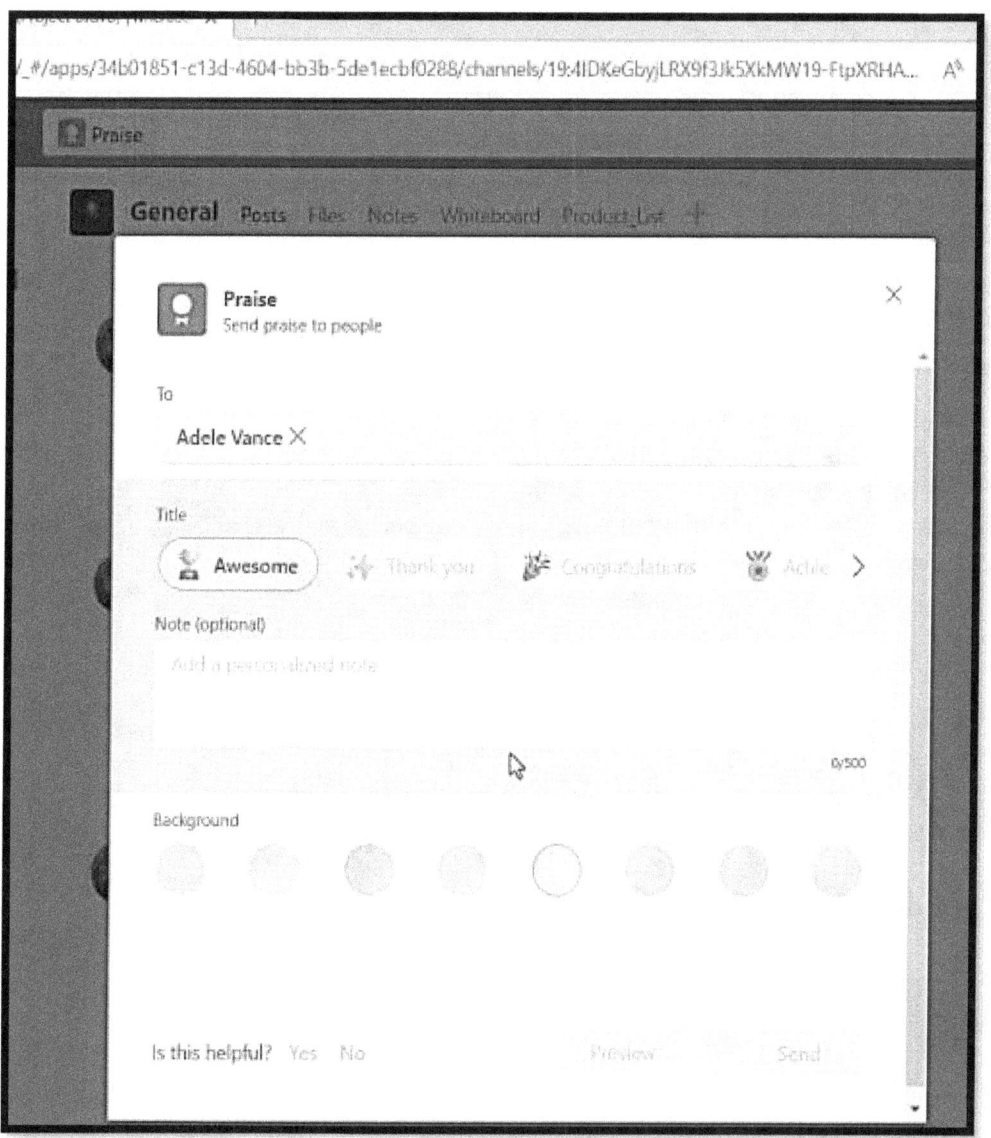

Another thing that's quite nice to do is if you type in the @ symbol again you could search for videos on YouTube and share those as well but bear in mind that if you don't have the YouTube app already installed in Microsoft Teams it will prompt you to do that. It's a simple case of clicking the "Add" button and then you can go in and search for the video you want to share. Then you can either copy the link or share it to a chat or you can choose to share in a meeting.

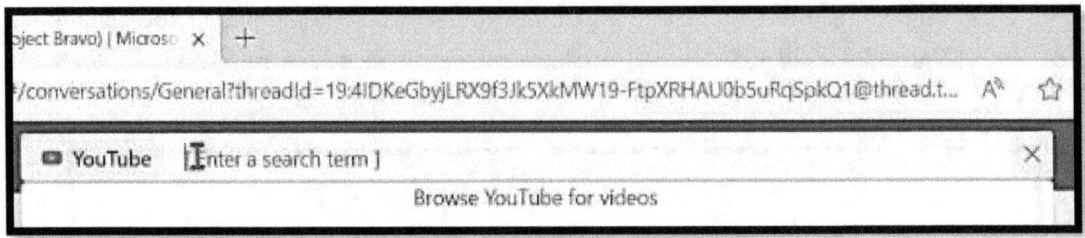

If you want to jump straight into a team meeting and share this video you can do that from here. Just click on "Join now" and say "Start sharing" and it's now sharing that video in your team's meeting.

Review Questions

1. Familiarize yourself with the basic navigation and features of the Microsoft Teams interface, such as the sidebar, chat, and calendar.
2. Practice using the keyboard shortcuts to quickly access common functions, such as starting a new chat or joining a meeting.
3. Customize the Teams app settings to align with your personal preferences, such as the theme, notification settings, and default app behavior.

CHAPTER 10
TROUBLESHOOTING TEAMS

In this **chapter**, we will be showing you how to fix Microsoft Teams not working. If you have any kind of problems with the Microsoft Teams app we will show you the quickest and easiest ways to fix it. In this chapter, we will be showing you how to fix it on an iOS device but you can do the same on Android as well.

Teams not working?

The first thing you should always do when something is wrong is to just restart your iPhone. If this does not help, try to reinstall the Microsoft Teams app: tap and hold on Microsoft Teams on your home screen and choose "Remove app" then choose "Delete app." Now reinstall it again from the App Store or Play Store.

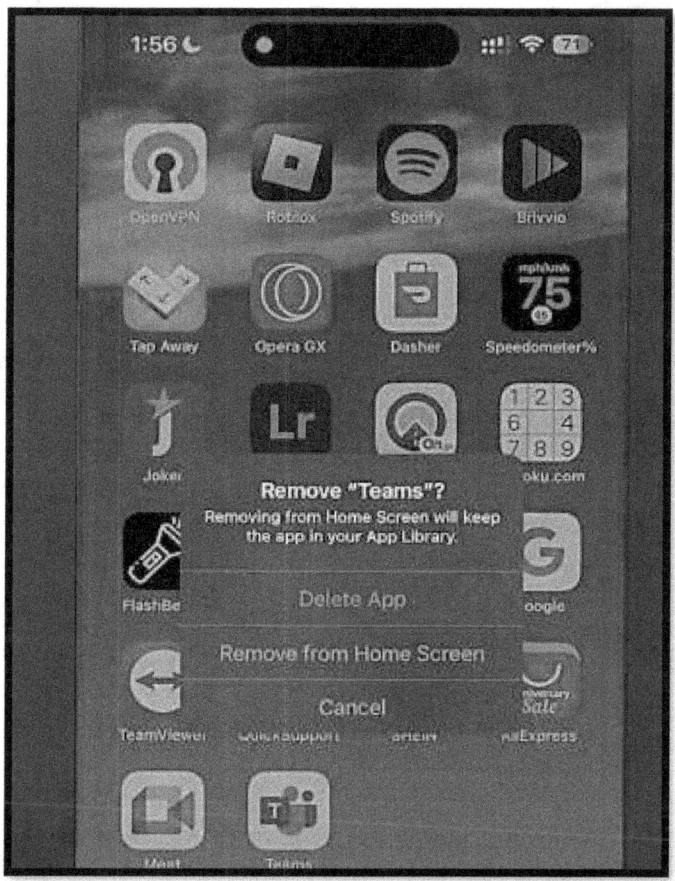

If this still does not help, try to switch your network connections. If you are on Wi-Fi try to use mobile data and if you are already using mobile data try to use Wi-Fi. The last thing you can do is to just reset your iPhone network settings. To do this, open Settings of your iPhone, go to General, and scroll down to "Transfer" or Reset iPhone, "now tap on Reset and choose" Reset network settings

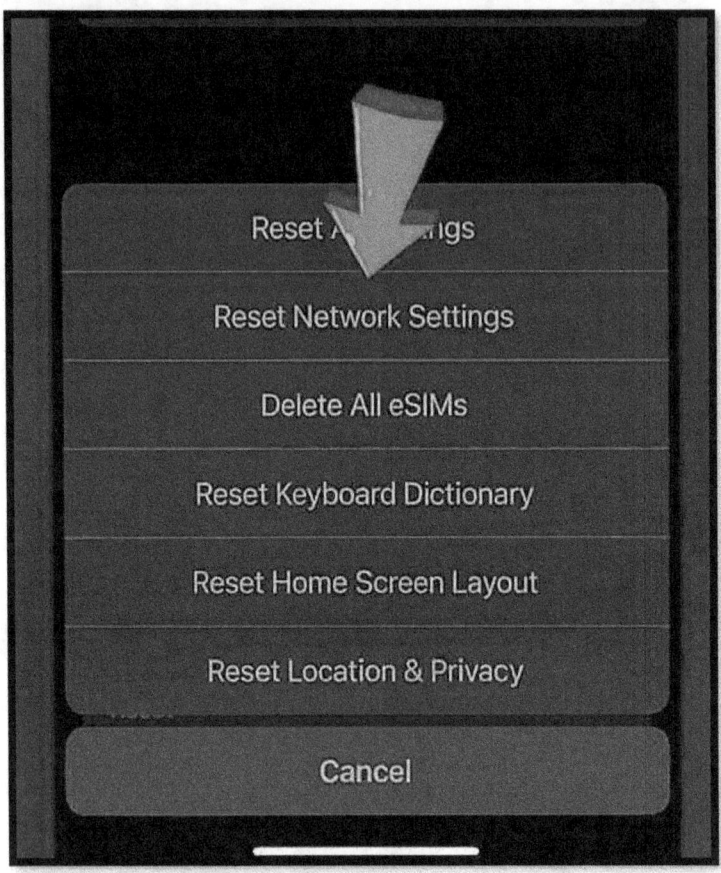

This will reset network settings and after your iPhone reboots your Microsoft Teams app should work properly without any problems.

Spell check not working?

In this section, we are going to see how to fix Microsoft Teams spell check not working problems. The first solution is to clear Teams cache files and disable and enable spell checks again. To do this, you just need to right-click on the start icon and select "Run window." Then you need to type "%appdata%" and click on okay. Here, simply double-click on the Microsoft folder, right-click on the Teams folder, and select the delete option.

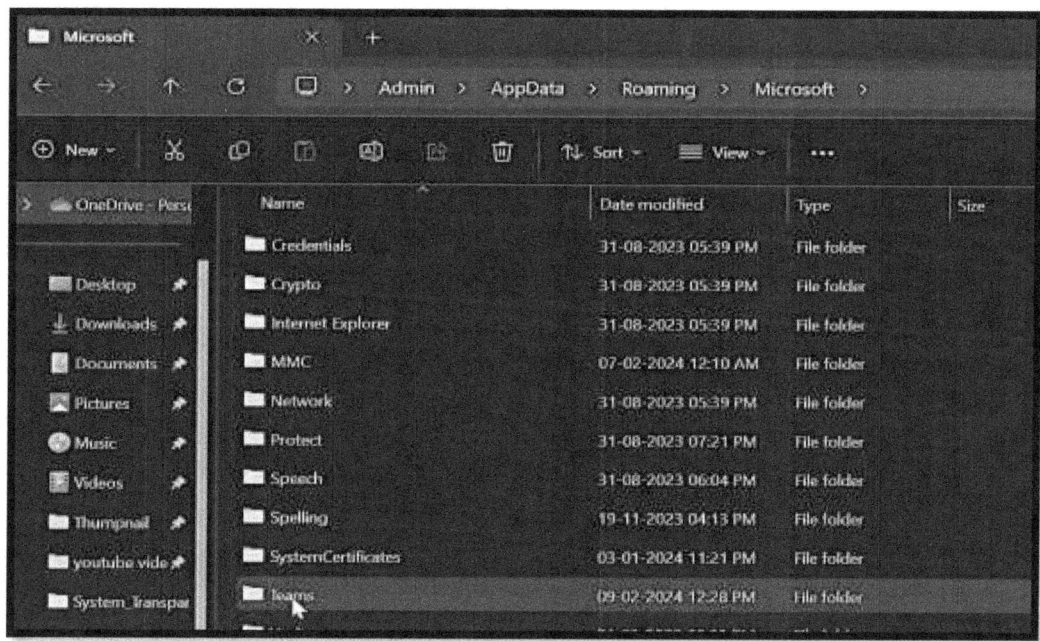

You need to delete this folder inside the Cache files of Teams. After reopening Teams this folder will be recreated. After you have cleared the cache file of Teams, open Teams, select the three-dot icon on the top, select Settings then uncheck spell check and close that.

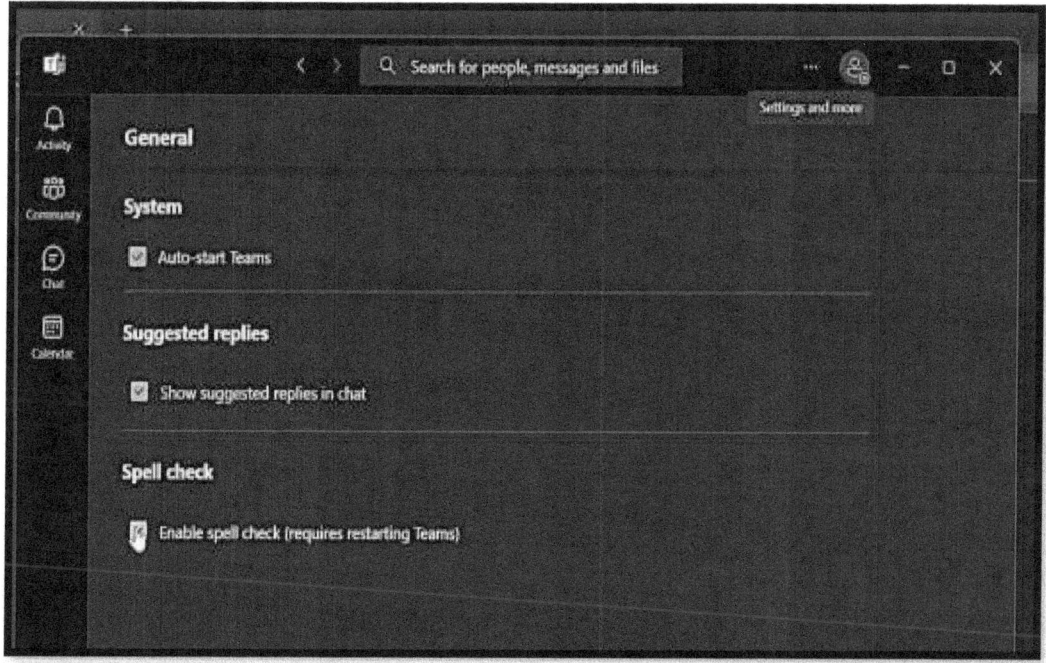

Now you just need to close in the background as well. Right-click on the Teams and select Quit. After doing this, open Teams again, now click on the three-dot icon on the top, select Settings, and enable the spell check option. After doing this, you just need to restart Teams and check if the problem is solved or not. If not, let's move on to the next solution. The second solution is to remove all languages other than English from your computer. The reason you'll do this is that Microsoft Teams uses the language of Windows when doing spell check and if there is more than one language in the Windows languages section, spell check may not work properly. To do this you just need to right-click on the Start icon, select Settings on the left side, then select Time and Language.

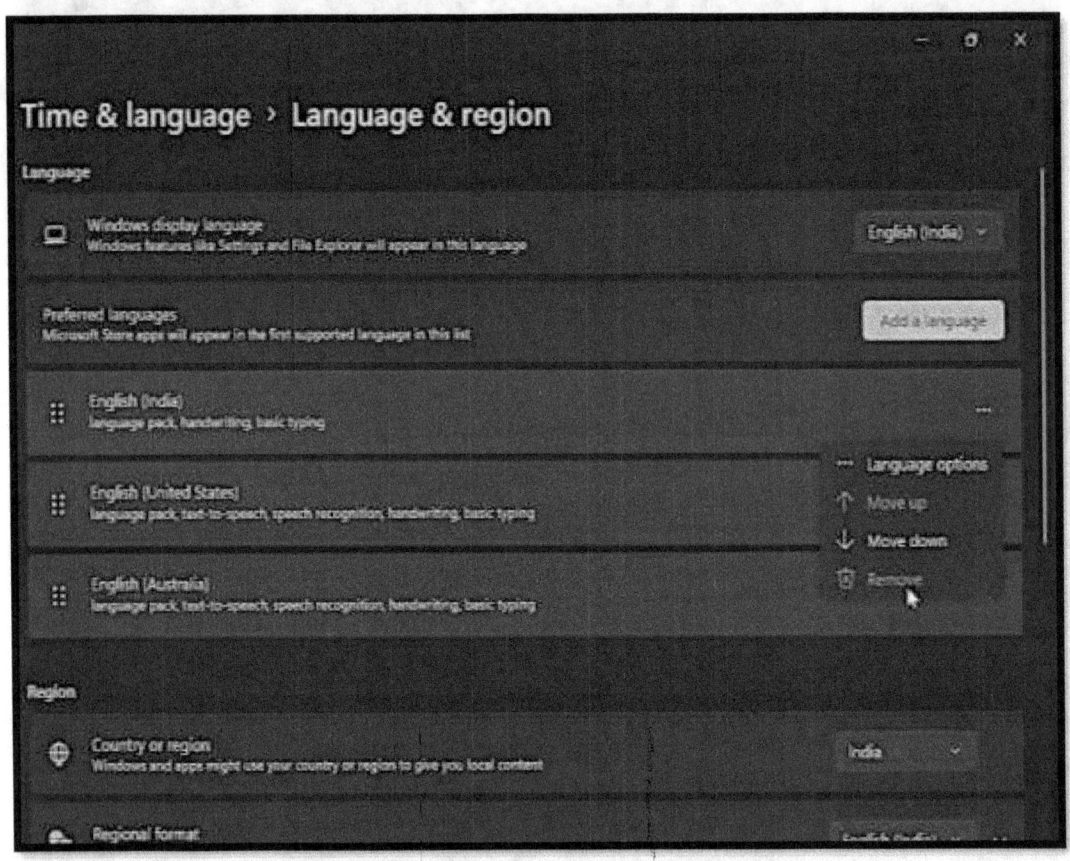

On the right side, select Language and Region and under the preferred languages remove the language other than English. After doing this, close this window then go ahead and restart your PC after which you can check if the problem is solved or not. If not, let's move on to the third solution. The third solution is to download the old version of Teams. The problem may be caused by the latest update and if this is the case downloading the old version will solve the problem. Before downloading you need to delete the Teams. Just go to the search menu and type Microsoft Teams, click on Uninstall, and select Yes.

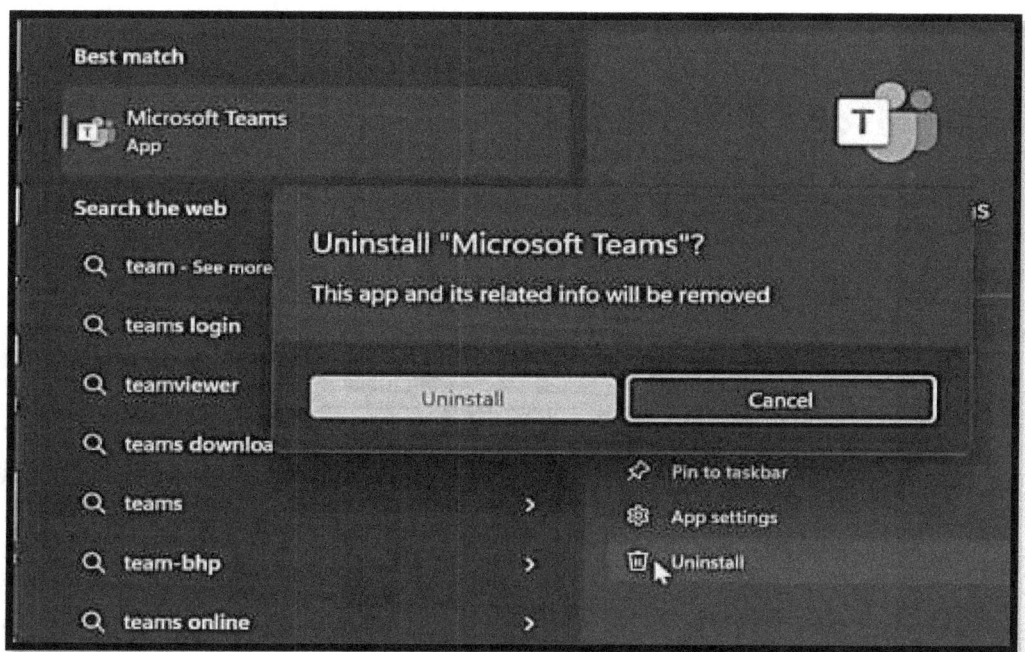

After uninstalling it, download and install the previous option and check if the problem is solved. One of these solutions will fix your problem.

Review Questions

1. Identify the common troubleshooting steps to take when the Microsoft Teams app is not functioning as expected.
2. Practice troubleshooting a specific issue, such as being unable to join a meeting or send a message, and document the steps taken to resolve the problem.
3. Discuss the potential causes of Teams-related issues.

CHAPTER 11
TEAMS APP AND ADD-ONS INTEGRATIONS

If you're new to Microsoft Teams you have the opportunity to add a bunch of apps and workflows inside of Microsoft Teams. If you want to connect something like Power BI or Google Analytics you can choose to connect apps inside of here.

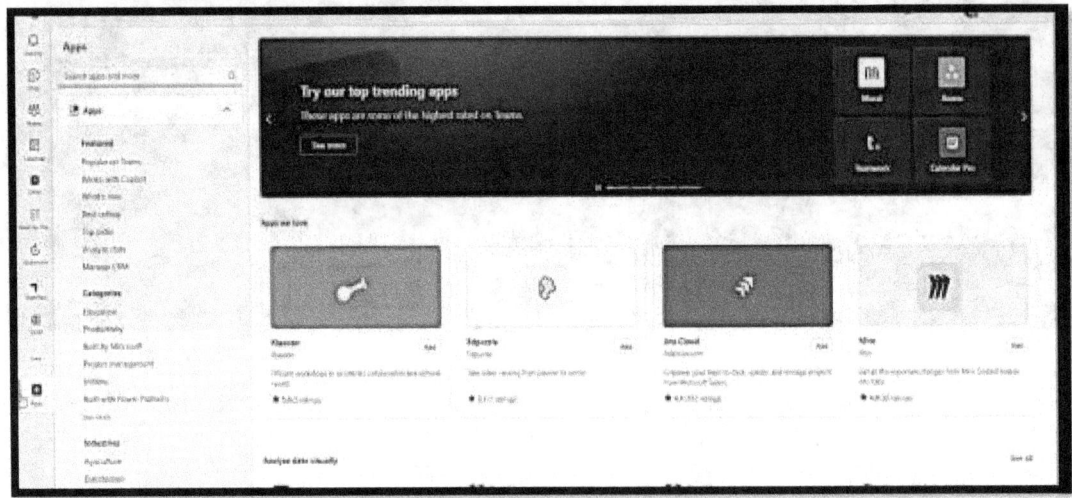

Note that the apps inside of Teams are so much to work with and it's only you and your team that can answer the question of what will work best for you. You've got a lot of integrations, and there are a lot of opportunities but just note that if you are using it for your team, you're new to Microsoft Teams and you don't know much about that app section but you're looking to integrate anything inside of your Teams, this area is where you will go to search for it.

App integrations

We're now going to look at some of the new features in Microsoft 365 that have made their way to Teams 2024.

Meet app

The first one is the Meet app. If you don't see that it'll be in your three-dot menu and you can search for Meet. You might need to add it in if it's not there by clicking the "Add" button and this is like a different view of your calendar so obviously you will have the Calendar app shown by default that gives you the normal view of your calendar. If you didn't know, you can jump into an agenda view which you might prefer instead to look a bit more like your phone app but the Meet app is supposed to bring out some richer items about your meetings. With the Meet app, you'll see your Teams meetings lined up and you can join your meeting right from here. You can reschedule it, you can chat with other participants straight from here and if you were getting close to the meeting it's got an "I'm running late button" so you can quickly message people to say "I'm running late" straight from Meet which is quite useful. Then underneath are your recently finished meetings. So rather than scrolling through the chat to find your other meetings or into the channel to find your meetings, Meet is bringing everything together wherever those meetings are from. If you pay for Teams Intelligent Recap via Microsoft Teams Premium or you're lucky enough to have Microsoft 365 Co-pilot then you'll be able to get Intelligent Recap via the Meet app. With this, you can quickly click on view recap, it then jumps into the Intelligent Recap bringing up the meeting notes that are generated by AI, and if you've got some tasks, you'll see that as well. Then on the front page of the Meet app, it also shows you where you shared any files so let's say we were using PowerPoint, we can jump straight to that PowerPoint slide if we want and the intelligent recap goes into action, picking out where it thinks there might be some tasks that we need to look at and where people have mentioned our names in this instance. Ensure to check out the Meet app, it's free and quite useful and you may find yourself using it more than a calendar to quickly join meetings and to quickly find your recaps rather than scrolling through the chat which might be taken up by some other things.

OneDrive app

The next new app is the OneDrive app which you might have seen change from Files. The app used to be called Files so if you've unpinned it and hidden it, again, that might be lurking in your three-dot menu. You might come into OneDrive thinking these files were yours that no one else could see since they're not in teams and you would be right. It shows you files from across the Microsoft ecosystem wherever they live so you can get back quickly to them which is useful if you understand the nuances about where to save things. As well as showing you all the latest files in OneDrive, Teams, SharePoint, or wherever they live in the Microsoft ecosystem you can quickly get back to them. If you set your default to open in the app then it'll ping out into Excel, Word, or PowerPoint for you. If you haven't done that you can just click on the three dots and change the default so you can either open it in a browser or an app or just open it directly in Teams depending on your preference.

Two new things that cropped up in the new OneDrive which are pretty useful and now more accessible in Teams are:

- Browsing files by people. You can see who you shared files with and then just get back to them.
- By meetings. Similar to the Meet app it's showing you the files associated with those meetings. If you want to go and watch the recording or the replay you can see all the past meetings there then just go and click on the files that you shared.

Planner

Planner is worth having a look at. To get things done across your organization we would encourage you to use Microsoft Planner. You can pin it into a tab of a team and that sidebar app then shows you all of the plans across every team that you're in and everything that's assigned to you. You can see quite quickly that just getting into Teams gets you into pretty much every app in the Microsoft 365 ecosystem which is why we usually start with Teams and that makes everything else easy. You can see everything assigned to you, that way, you never miss a task and the planner chases people up for you so if you assign a task to someone and provide a due date it's going to chase them up multiple times before the due date is done and then after they don't do it and miss their deadline it will keep chasing them until they do it. Ensure to make use of Tasks by Planner which is now about to be called Microsoft Planner.

Viva insights

Again, this one might be one that's not defaulted so it'll be in your three-dot menu. With this, you can get access to a lot of things for free.

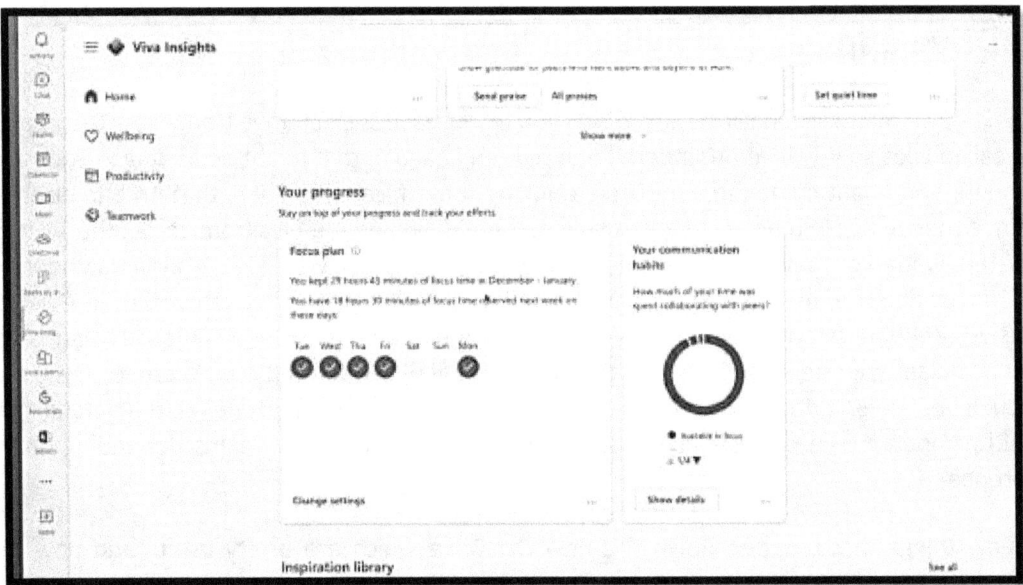

It's got Lo-Fi times, Zoom out, and Focus that's recommended for us right now. You can get some insights about your meetings but the most useful thing for Microsoft Teams that most people don't turn on is Focus Plan. If you've got a focus plan already set up, you can set some lunch hours for you to block out time for lunch but if you tend not to have lunch and you prefer to work straight through you don't need to have that turned on.

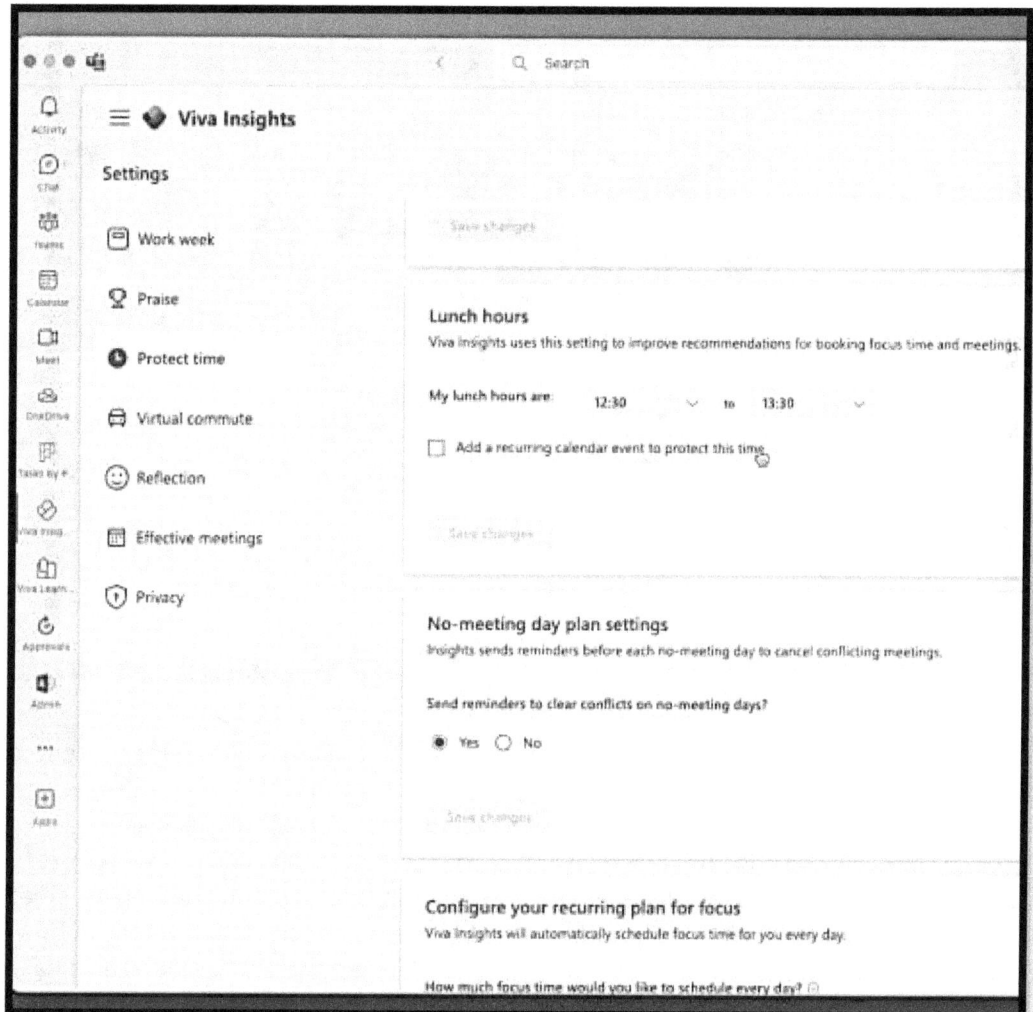

You can have no meeting days turned on if you want to but you have to pay an extra license for that. However, what you can do for free is manage your own Focus time, that way, you have big blocks of time to focus and get things done. You can also get it to silence notifications for you from Teams while you're in that Focus time and if you set all that and say what days you want Focus time on, it'll try and block that time out in your calendar.

SharePoint

The last thing that isn't new but a lot of people miss is that when you set up a team, it sets up quite a lot of things in the background for you as well as a fully-fledged SharePoint site which is where it handles all the documents but also handles a lot of the pages so you've got a fully-fledged internet.

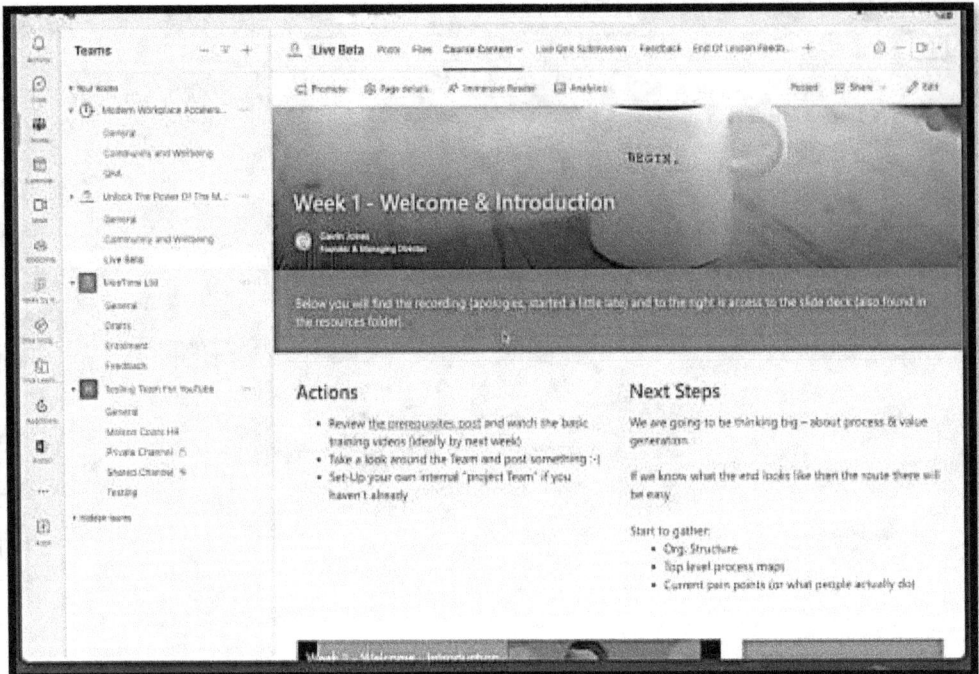

It's as powerful as a full company internet in each team so you can use that to bring content to life even within a team and everyone in the team's got the ability to make any pages they want. You could make a little course content and outline with multiple web pages, steps and action, and links to videos and links back to documents and just bring your content to life. You can also use external links and embedded forms to make meeting minutes or outputs for meetings or prep for meetings or whatever you want to collaborate on. Just bring it to life with a bit more content in how you would consume it in your personal life so rather than attaching a PowerPoint slide to an email, you can now bring and start to put out an internet linked web pages with pictures, videos, or rich content; just think about different ways that you would like to consume content and people that you're collaborating with might want to consume content and put it into a SharePoint page within the team. You can get one pinned into your channel and everyone can go through that and add to that. That way, you start to build out a knowledge base in your team and it's just really powerful. SharePoint pages can do a lot of things for you. It has been since Team's inception but then a lot of people miss this.

Review Questions

1. Go to the Microsoft Teams app store and identify 3 third-party applications that can be integrated with the Teams environment.
2. Practice installing and configuring the selected add-ons and observe how they integrate with the Teams interface.
3. Discuss the benefits of integrating applications with Microsoft Teams.

INDEX

"

"New chat", 49
"Save this message", 46, 47

3

365 group, 26, 89, 92

A

A code is something that can be given to you by the team owner, 30
A good option, 50
a little bit more information, 1
a task management tool, 2
ability to annotate a PDF in an assignment, 95
ability to customize, 14
ability to make any pages they want, 128
about improvements, 21
access Breakout Rooms, 77
access Teams, 6
access the files, 18
access to your settings., 52
Accessibility, 53, 61
actual team, 26
add a description, 27, 34, 35
add a description,, 34
add a GIF, 43
add a triangle, 74
Add Channel, 35
add other tabs, 49
add photos, 83
add text boxes, 73
add the PDF, 96
add things, 14, 34, 73, 115
Add to My Day, 88
adding a picture, 56
adding apps, 115

Adding fun to your chats, 42
Adding new tabs, 113
Addressing a specific person, 45
Addressing common issues is very important, 2
addressing specific people, 2
advance of the meeting, 76
Advanced meeting scenarios, 77
align with the project's lifecycle, 84
align with your personal preferences, 118
allow members, 34
allowing users to chat, 3
allowing users to enhance their experience, 5
allowing users to schedule and join meetings, 5
an @mention, 33, 34
an application, 68, 72
an appointment, 63
an organization-wide team, 26
analytics of your teams, 23
Analytics section, 34
Annotate a PDF, 95
annotate and add other elements, 73
annotating PDFs for educators, 95
annotation, 2, 70, 108
announcements, 83
Announcements, 28
another application, 10
any kind of problems with the Microsoft Teams app, 119
anyone in your organization, 27
App ecosystem, 5
App integrations, 124
app is not functioning as expected, 123
appdata, 120
applications, 7, 22, 28, 30, 46, 83, 128
approvals, 83
assign appropriate roles and permissions., 84
Assign approvers, 83
assign participants, 78
Assign participants, 78
assignment instructions, 105
attach files, 42, 58

audio calls, 49, 52
automatically integrates the Edge Browser with Teams chat., 20
automatically popped right into the to-do list, 16
Availability, 6

B

background, 56, 61, 101, 102, 103, 117, 122, 127
background noise suppression, 103
Background Noise Suppression button, 103
Before you initiate video and audio calls, 52
better search, 2
big difference, 1
big long list, 6
bigger reply window, 43
bit more content, 128
bit untidy, 56
blank canvas, 72, 74, 75, 76
blur out the background, 56
Bookmarking, 46
bookmarking important conversations., 2
Bookmarking, editing, and deleting posts, 46
boost someone's morale, 117
bottom right-hand corner, 42
brainstorm different ideas, 74
brainstorm ideas, 76, 113
Brainstorming ideas, 72
Brainstorming ideas in a meeting, 72
brainstorming sessions, 2
brainstorms ideas and projects, 72
brand new homepage, 19
brand new window, 112
brand-new OneDrive, 18
Breakout Room, 77
breakout rooms, 77, 78, 79, 80
Breakout rooms are a great way for a meeting organizer, 77
Breakout Rooms work, 77
bring everybody back to the main meeting., 79
browse by meetings, 18
browser integration, 2
Browser integration, 20
browsers and mobile apps, 6
Browsing files by people, 125

budgets, 83
build out a sample project channel, 84
bunch of apps and workflows, 124
bunch of new caption settings, 21
bunch of shortcut links, 116
business subscriptions of Microsoft 365, 77
bypass the lobby, 11, 64

C

cache files, 120
Cache files of Teams, 121
Calendar app, 124
calendar open, 68
calendars, 5, 81
captions are going right across the top, 22
categorize and organize related project documents, 83
centralize work happening across a variety of platforms, 83
centralized hub for teamwork, 3
chance to be in the meeting, 11
change the color of the font, 44
change the language if you want, 21
change your view, 60
changing shortly, 110
Channel or a chat, 16
Channel or an email, 30
channel settings, 34
channels inside of Teams, 15
Chat, 5, 8, 28, 48, 58, 110
chat, and calendar., 118
Chats are conversations, 39
chatting in a team Channel, 58
check before setting up a Teams meeting, 54
check for meetings, 20
check shared files, 20
checking a few settings, 52
choose "Delete app.", 119
choose a file, 69
choose a page, 100
choose a task list, 14
choose different colors, 21
choose to manage your teams, 23
choose to share the Outlook window, 68
choose Windows, 68

130

choose your guardian account, 106
Choose your preferred language, 21
clear folder structure, 83
click "Remove from My Day", 86
click "Start a post", 109
click Apps, 10
click on "Download Microsoft Teams Desktop, 3
click on the app launcher, 72
click on the Open button, 4
Click on the Rooms option, 77
click on the three dots, 37
click Pending invites, 34
Click Settings here, 21
clicking "Copy Component", 15
clicking on the link in the body of the message, 66
clicking the "New" drop-down, 9
clicking the join button, 27
co-authoring documents, 72
collaborate on various projects and tasks, 3
Collaborate with Loop, 14
collaborating, 2, 52, 72, 128
collaboration platform developed by Microsoft, 3
collaboration within organizations, 3
Collaborations, 5
collaborative, 14, 77, 108
Collaborative, 2, 14, 22
collaborative Loop, 14
Collaborative Loop, 2, 14, 22
collaborative Loop components, 14
colleagues and team, 72
colleagues or external participants, 5
collection of useful tips and tricks, 2
color different aspects, 101
coloring book, 100, 101
combining Planner, 85
come off mute, 62
come to manage your tags, 34
comments, 73, 115
communicate more easily, 1
Communication, 82
communication clearer, 1
communication needs., 84
Communications hub for your project, 82
complete guide, 2
completed section, 87

Components are now supported in channels, 14
components to planner, 16
composing a message, 44
confirm the language everyone is speaking, 58
confirmation, 98
connect with other tools, 5
connect with other tools they use in their workflows., 5
connection icon, 107
consume content, 128
consume content and people, 128
contact details, 32
content, 82, 111, 112, 128
contracts, 83
conversation field, 42
conversation thread, 110
conversations, 2, 41, 42, 43, 49, 50, 52, 58, 82, 83
cool little mode, 60
co-organizers, 11
copy components, 15
copy the meeting link, 58
correctly using teams, 82
couple of little things, 115
couple of students, 99
course content and outline, 128
covering all the new features and tips, 1
create a new channel, 7
Create a new channel, 38
Create a new channel within an existing team, 38
create a new chat, 39, 49
create a new tab, 115
Create a new team, 38, 84
Create a new team using a predefined template, 38
create a nice visual effect, 86
create a private team, 23
create a public team, 23, 27
create a tag called Project Leads, 34
Create a task from a post, 110
create a task in the planner, 110
create a team, 24, 25, 26, 28, 35, 82
Create a team, 24, 37
create dedicated channels, 82
create new chats, 112
create new tags, 34
create private channels, 32, 34, 35

create some Breakout Rooms, 77
create tags, 2
create Teams meetings, 67
create the tab, 114
create the team, 27, 29, 30
create things like New channels, 35
create two to four rooms, 77
creates a new entire window, 40
creating a program team, 82
creating a public team, 29
Creating a team, 24, 26
creating a team based on that group, 25
creating all of the channels, 30
Creating Tags, 34
creating tasks from posts, 2
creating teams, 31
CREATING TEAMS, 23
CREATING TEAMS AND CHANNELS, 23
critical announcement, 83
currently working on, 7
custom-built apps, 83
Customize a default set of reactions, 17
customize all of the settings, 32
customize channel settings, 2
Customize default reactions, 17
Customize the meeting details, 81
Customize the plan views to suit your preferences., 94
Customize the Teams app settings, 118
customize your behavior, 53
customize your language, 61
customize your language and speech, 61
customize your settings, 64
customizing pop-up windows., 2

D

date or version numbers, 83
decide to add a participant, 58
default app behavior, 118
default Avatar, 32
default General channel, 26
default reactions, 17
delete and restore channels, 32
Delete team, 37, 38
delete teams, 37

delete the message, 41
Delete the team, 37
deleting, 33, 38, 46
Deleting a team, 37
delivery, 82
demonstration in Outlook, 68
department, 82
desktop app, 13
desktop application, 6, 77
desktop application for Windows, 6
desktop version, 77, 81
details of the meeting, 12
device or location, 83
Device settings, 61
difference between choosing a window to share, 68
different components of Planner, 2
different features, 22
different items, 31
different options, 14, 22, 35, 44, 58, 62, 67, 100
different parts of Teams, 2, 6
different picture, 32
different PowerPoint slides, 70
different project or team-based plans., 94
different seats, 60
different tabs, 7, 32, 49
different tabs running across the top., 7, 32, 49
different teams, 11, 23
different things, 34, 50, 67
digital activity your student, 108
direct communication, 34, 39
Direct linking, 16
direct linking from task lists, 2
disable certain permissions, 32
discover the power of Microsoft Planner, 2
Discuss, 108, 123, 128
disguise your background, 56
do a quick post, 14
Do Not Disturb, 69
document review, 108
document the process, 84
double-click to create a new event you can create, 67
Download app for desktop, 3
download in your web browser, 3
Download Teams for home, 3
download Teams for work or school, 3

Downloading, 3
Downloading and Installing, 3
downloading the old version, 122
draw your attention, 86
drop-down, 35, 55, 57, 62, 63, 87, 91, 116

E

each team and each channel, 7
Easy way to find channels or chats, 16
easy-to-use task management solution, 85
Edit mode, 48
edit the Whiteboard, 76
editing, 46, 76, 104
educational institutions,, 3
educational setting, 108
educators and students, 2, 95
Effective communication is key, 2
effectively co-authoring, 70
embedded forms, 128
emojis, 33, 42, 51, 58
Employee Onboarding, 30
enabling users to create, 5
encourage using descriptive names, 83
enter the date and time, 98
Enterprises and school organizations, 19
entire portfolios, 82
entire screen, 68
entire team, 33
establishing your organizational structure, 82
everybody moved online, 52
everyday work, 1
everything from the Wikipedia site, 20
exact link, 41
example project of IT infrastructure, 82
Excel, 5, 18, 69, 71, 72, 92, 112, 115, 125
Excel inside of Teams, 112
Excel Live., 69
exclamation icon, 83
executing actions, 7
expiration dates, 83
explore the different calendar views, 13
explore the interface, 6
explore the key features of Microsoft, 2

Explore the options for sharing files and other content within a chat, 51
EXPLORING NEW FEATURES, 14, 95
EXPLORING NEW FEATURES IN TEAMS, 14
Extend due date, 98
extend our experience, 10
extensibility, 83
external links, 128

F

facilitate communication, 2, 3
Familiarize yourself with the basic navigation and features, 118
feature provides an overview of your involvement, 94
feedback, 2, 82, 83, 99, 108
feedback for official approvals, 83
feeling of togetherness, 60
few different choices, 24
few different methods, 31
few little tricks, 11
few other options, 40
File Explorer, 69
file library, 9
Files app, 18
Files area, 9
final option to delete a team, 36
finding things, 11
focus on sharing a screen, 67
font size, 21, 22
formatting, 40, 43, 44, 51, 73, 83
Formatting your chats, 43
four different meeting rooms, 77
fourth new feature, 17, 99
free app in Microsoft 365, 16
From a group, 26
From scratch, 26
From template, 28
frustrated, 100
full company internet, 128
fully-fledged internet, 127
further settings, 61

G

Gallery format, 59
General, 26, 28, 42, 56, 63, 93, 112, 120
general Channel, 7, 23, 41, 113
general Channel by default, 7
general channel in a team, 31
general conversations, 7
get a notification, 48, 51
get a preview, 21
get Intelligent Recap via the Meet app., 125
get pinged about your post, 109
get the most out of this book, 1
gifs, 5, 33, 42, 58
giving your messages a bit of personality, 43
going on across your teams, 9
good idea, 54
goto option, 116
greatly improve collaboration and productivity, 2
group and notifications tabs, 93
group discussion, 77
groups of four to work on a question, 77
guests, 26, 32, 33

H

harmonizes task management, 85
having many workshops, 77
having your camera off, 61
height, 21, 22
hidden team section, 37
hide a team, 36, 37
Hiding a team, 37
hosting webinars, 53
how to create and manage Teams and Channels, 2
How to set your pop-up windows, 112
how to structure projects, 2
huge lifesaver, 84

I

identify 3 third-party applications, 128
images, 42, 56, 74, 115
Imagine breakout rooms, 77

important for effective communication and organization, 2
important project documents, 83
Important settings, 92
improve your work with others, 1
improvements to live captions, 2, 21
including a Microsoft team., 89
including the Teams Calendar, 2
incoming calls, 111
indicate basic, 89
individual projects, 82
individual whiteboards, 76
information for your child., 107
information you can choose to send the meeting invites, 12
infrastructure program, 82
Initiate a call on demand, 56
initiating an on-demand meeting, 56, 62
initiating on-demand calls, 2
inside Microsoft Teams, 112
inside of Enterprises and school organizations and Stream, 19
inside of Microsoft Teams, 124
Insights bottom, 108
instead of selecting one of the existing, 76
integrate your chats, 20
integration, 5, 83
integration allows for easy access to files, 5
Integration of external tools, 83
Integration with Microsoft 365, 5
integrations would include Trello and Asana, 83
Intelligent Recap, 125
interesting site, 20
internet linked web pages, 128
interrupting the flow of the presentation, 58
INTRODUCTION, 1
invite participants, 77
IOS and Android devices, 6
iOS device, 119
iPhone or Android, 106
It has a clean and simple navigation menu,, 85

J

Join or Create, 26, 30, 37

Join or create a team, 23, 30
Join room', 79
Join team, 30, 37
joining a meeting, 118
Joining a scheduled meeting, 65
Joining a team, 30
joining public and private teams, 31
jump into a meeting to discuss, 56
jump straight, 118, 125

K

keep discussions organized, 5
Key features, 5
key on the keyboard, 42
key takeaway, 81
keyboard shortcut, 59, 67
keyboard shortcuts, 11, 118
kinds of fun things, 42

L

Language and Speech, 21
large enterprises., 3
learn helpful information and tools, 1
learning how to use Microsoft Teams, 1
learning how to use Microsoft Teams effectively, 1
Leave or End meeting., 62
Leave team, 37
Leaving a team, 36
Let's talk about adding new tabs at the top, 113
let's talk about how we can delete messages, 48
Let's talk about leaving a team, 36
letting the other people into the meeting., 11
leverage Teams, 82
list checklist, 14
Lists from Microsoft To-do, 90
little bit of time, 26
little comment icon, 73
little contextual menu, 7
little course content, 128
little heart, 59
little message, 34, 48
little message board, 34
little parent, 107

little pop-up notification, 42
little pop-up window, 32
little prompt, 103
little red icon, 69
little text box down the bottom, 109
Live are two great additions, 72
Live captions improvements, 21
Live events, 53
live presentations, 2, 69
Lo-Fi times, 126
look at your profile picture, 69
looking at messages, 48
looking for conversations, 50
looking for your files., 45
Loop component, 16
Loop components in channels, 14
Loop page, 85

M

main teams list., 37
major benefit of teams, 83
make audio and video calls, 3
make changes, 21, 48, 76
make diagrams, 74
make teamwork smoother, 1
make work easier and more productive for your team, 1
make workflows and team collaboration even smoother, 2
make your first component, 14
make your teamwork smoother, 1
making the necessary settings, 79
manage apps, 40
manage participants into different groups, 77
manage projects better, 1
Manage team, 26, 32, 34, 35
Manage Team, 30
manage team members, 2
manage the tasks, 85
managing schedules, 2
Managing teams, 32
Managing Teams, 31
managing teams and channels, 31
managing your teams and your channels, 7

Mark as important, 83
Mark Team notifications, 17
Marketing team, 6
marketing-type posts, 6
master project management from end to end., 82
Meet app, 2, 18, 19, 124, 125
Meet button, 55, 56
Meet now, 11, 55, 56, 62
meeting can bypass the lobby, 64
meeting for a review later, 60
meeting information, 65, 66
meeting minutes, 128
Meeting options, 61
meeting organizer, 53, 77, 78, 79, 80
meeting participants., 69, 71, 72, 76
meeting whiteboard Tab, 65
meeting window, 79
Meetings, 5, 53
member link in the top corner, 32
member status, 32
memes, 33, 34
message button underneath, 43
microphones on mute, 53
microphones turned on, 64
Microsoft 365 homepage, 72
Microsoft 365 on the homepage, 20
Microsoft Edge, 68
Microsoft Edge browser, 68
Microsoft Stream is our video tool, 19
Microsoft Teams app store, 128
MICROSOFT TEAMS FOR EDUCATION, 95
Microsoft Teams has become the central hub for managing projects, 82
Microsoft Teams on your home screen, 119
middle of the lower part, 107
mobile data, 120
mobile phone, 106
more accessible in Teams, 125
more emailing files, 83
more of team announcements, 34
More Options section, 110
more participants, 81
more to boost collaboration, 82
Moving over to Flagged emails, 88
Multiple team members, 5

Multiple team members can collaborate simultaneously on documents, 5
multiple web pages, 128
Music and Movie, 30
Music and Movie recommendations, 30
muting irrelevant conversations, 2
My Day, 85, 86, 88
My Day in Microsoft To-do, 85
My Plans, 88, 90, 92, 94
My Tasks, 87, 88
My Teams, 91
My Teams look very similar, 91

N

New assignment, 95, 104
New conversation, 42
new conversation button, 7
new features, 1, 2, 14, 95, 124
New meeting schedule, 11
New meetings, 11
New Plan button, 90
new practice words, 108
nice non-verbal way, 59
noise suppression, 2, 102, 103
Noise suppression for reading progress, 102
note about chats and typing a message, 40
notes, 70, 73, 74, 86, 113, 115, 125
notification, 30, 47, 50, 98, 111, 113, 118
notification settings, 111

O

Observe how the annotations are displayed, 108
observe how they integrate with the Teams interface, 128
Observe the meeting invitation, 81
Old vs. new planner, 92
Onboard Employees, 28
one language in the Windows languages section, 122
one member from that channel, 8
OneDrive, 2, 5, 14, 18, 44, 69, 125
OneDrive app, 18, 125
OneDrive thinking, 125
one-on-one or group chats, 5

open a calendar invite, 63
open Settings of your iPhone, 120
open Teams, 121, 122
Open the "Calendar" section, 13
Open the Teams calendar and practice scheduling a new meeting, 81
open up an assignment, 98, 102, 105
open up the to-do list, 16
Opening and Exploring Teams, 6
option and type the code, 30
option at the bottom, 37
option for an agenda, 12
option for students, 104
option to join with an ID, 11
option to view that recap, 18
options for sharing files, 51
options to create Windows, 112
organization, 23, 26, 27, 30, 49, 109, 113, 126
organization-wide group, 27
organizer or presenter, 59
organizers, 11
other content, 51
other files, 18
other resources within the Teams' interface., 5
other teams and channels, 44
Outlook, 5, 14, 25, 26, 63, 66, 67, 68, 88
outputs for meetings, 128
OVERVIEW, 2
own TPS report, 95
owners and members, 32

P

participants can edit, 76
participants' pane, 58
password protection, 83
PC straight, 9
pending invites, 23
people are active inside of your team, 34
people pane, 59
personal plan, 85, 90
pictures, 128
plan feature, 85
Plan Settings, 93
Planner, 2, 10, 16, 85, 87, 88, 90, 91, 92, 93, 94, 126

planner board, 110
Planner board, 16
popular collaboration platform, 2
pop-up notification, 43
portfolio container, 82
position, 21, 73
positioning at the bottom, 22
posting and receiving messages, 41
Posting and receiving messages, 41
potential causes of Teams-related issues., 123
Power BI or Google Analytics, 124
PowerPoint, 5, 18, 69, 70, 72, 125, 128
PowerPoint Live, 69
PowerPoint slide, 125, 128
Practice installing and configuring the selected add-ons, 128
Practice troubleshooting, 123
Practice troubleshooting a specific issue, 123
prep for meetings, 128
presentations, 9, 72, 83
presenting Teams, 69
press CTRL + SHIFT+ K, 59
prevent confusion, 83
previews on the notification, 111
private chats, 8, 49, 116
private group, 27
private messages, 48, 49
private or public, 34
private plan, 91
private team, 23, 27, 30
productivity, 82
program team, 82
programs, 82
project management, 2, 82, 83, 84, 89
promote or demote, 32
provide a structured way to collaborate, 5
public team, 23, 26, 27, 29, 30, 35, 37

Q

Q&A, 14, 65
questions, 82, 83
quick access, 18, 83
quick and impromptu meeting, 11
quick message, 32, 116

quick title and instructions, 95
quickest and easiest ways, 119

R

reaction tray, 14
reading assignment your student, 102
reading progress, 102, 108
reading progress is one of our learning accelerators, 102
recent meetings, 18
recording options, 5
Reflect feature, 100, 101, 102
Reflect for grown-ups, 101
Reflect for staff, 101
Reflect improvements, 95
Reflect mindful coloring book, 100
Rejoining a team you left, 37
remember to set permissions, 83
reminders, 2, 19, 53, 98
Reminders, 98
Remove app, 119
remove apps,, 34
Reset iPhone, 120
Reset network settings, 120
reset your iPhone network, 120
resolving problems, 2
rich content, 128
right people get access to the right channels, 82
right-click on the Teams folder, 120
roles changed within the team., 38
Room's icon at the top, 77
Run window, 120

S

Sales and Marketing, 6
Sales and Marketing is a team, 6
Sales and Marketing team, 6
sales assistant, 34
same OneDrive, 18
schedule a meeting, 55, 63, 65
Schedule a meeting, 56, 63
scheduled meeting, 65
scheduling and joining meetings, 2

Scheduling Assistant, 63
scheduling meetings, 67
school connection app, 106, 107
School connection app, 106
school data sync,, 106
screen sharing, 2, 5
search bar at the top, 32, 115
second new feature, 16, 98
second new feature is also from Loop, 16
section for captions and transcripts, 54
see a couple of options, 98
see a little school, 107
see an overview or some statistics, 33
see the direct linking, 16
select "Run window.", 120
select a different due date, 98
select Ask for approval, 83
select Settings, 121, 122
select Teams, 6
select that student in the list, 98
select the delete option., 120
select the three-dot icon on the top, 121
select Time and Language, 122
send a message, 33, 42, 53, 116, 123
send messages, 42
send people a reminder to respond, 19
send personal invites to everybody, 11
Send Reminder, 19
Sending private messages, 48
separate application, 72
separate rooms, 79
series of checkboxes, 32
set up a tag for sales assistant, 34
set up your audio, 56
settings and modify, 52
settings available in the Teams app, 13
settings icon, 76
Settings tab, 30, 34
SETUP MEETINGS AND CALLS, 52
Share a Microsoft whiteboard, 76
share a Whiteboard with other meeting participants, 76
share files, 2, 3, 5, 44, 72, 83
share your ideas., 76
share your Outlook window, 68

shared Channel, 34
shared files, 83, 125
shared locations, 15
Shared Plans, 91
Shared With, 91
SharePoint, 2, 5, 35, 83, 109, 125, 127, 128
SharePoint page, 128
SharePoint page within the team, 128
SharePoint pages, 128
Sharing and accessing files via chat, 44
Sharing files, 83
sharing files,, 52
Sharing your screen, 67
show captions in your meetings, 53
show more information, 108
showing you how to download, 2
showing you the files, 125
sign in with your Microsoft account, 5
significant popularity, 6
similar name, 82
simply double-click on the Microsoft folder, 120
single click, 15, 17
small businesses, 3
smaller or medium,, 22
some additional tabs, 65
some of the key features, 5
some of the new apps appearing in Microsoft for 2024, 109
somebody creates a public team, 30
speaker, 57, 58, 59
speakers, 57
specific about certain topics, 7
specific Channel, 33
specific conversations, 6
specific ID, 11
specific person, 110
specific team, 17, 48, 116
specifications, 83
Spell check not working?, 120
spreadsheet, 71, 72, 115
staff account owner, 102
stakeholders in your post, 83
stakeholders outside, 83
standalone microphone, 57
start a chat with a member, 32

Start a post, 14, 109
start a video call, 32, 49
start adding tasks, 14
Start by creating a team, 82
start clicking on the color, 101
start having conversations, 26, 28
start the recorder, 64
start to build out a knowledge base, 128
start typing, 27, 40, 73, 74, 109
start typing in the name of somebody, 27
Starting a meeting, 55
starting a new chat, 112, 118
Starting with Shared plans, 89
Starting with structuring teams, 82
Starting with the Private task, 88
status updates, 82, 83
Staying up-to-date with the latest tools, 1
stereo microphone, 57
stickers, 5, 33, 34, 42, 58
sticky note-style graphic, 73
Stop sharing, 69, 72, 76
structure, 2, 84, 109
Structuring, 82
struggling with an exercise in room, 79
student has an extended due date, 99
students' version, 100
subject matter, 26
super simple task, 41

T

tables, 14, 44, 105
Tags allow you to quickly reach a group of people, 34
take a look at some of the basics, 6
Take control, 70
talking about initiating video, 52
Tap that then load up the school connection, 107
task assignment and commenting, 22
task list, 14, 16, 85
tasks, 2, 82, 85, 86, 87, 88, 110, 116, 125
teaching you how to use all the different parts of Teams with ease., 2
team channel, 7, 48
Team channels, 6
team communication, 34

team meeting, 55, 62, 64, 66, 72, 118
team participants, 72
team tasks, 10
team's icon, 85
TEAMS APP AND ADD-ONS INTEGRATIONS, 124
Teams are very interesting, 23
team's **calendar**, 10
Teams Calendar, 11
team's channel, 49, 56, 64
team's Channel, 8, 58
Teams channel or chat, 108
Teams Chats, 2, 39
TEAMS CHATS, 39
TEAMS CHATS AND CONVERSATIONS, 39
Teams environment, 128
team's files library, 9
Team's inception, 128
Teams integrates with other Microsoft Office applications, 5
Teams is not immediately obvious, 46
Teams is now used heavily for video and audio meetings,, 52
teams listed, 23
Teams meeting, 67
Teams meetings, 125
Teams' meetings, 53
Teams' mobile consumer app, 106
Teams not working or spell check not functioning., 2
Teams not working?, 119
Teams offer audio and video conferencing capabilities, 5
Teams organize conversations, 5
Teams seamlessly integrate, 5
Teams seamlessly integrate with other Microsoft 365 services, 5
text, 39, 73, 82, 109, 115
text message., 39
the "Join now" button, 58
the "Leave" button, 62
the "More actions" button, 60
the "New Whiteboard" button, 76
The activity center, 110
The **Activity** option, 9
the App Store or Play Store., 119
the basic navigation, 118

The basics, 109
the chat and the context, 20
the consolidated way, 14
the context of the chat, 20
the conversations, 50
the desktop version of Teams, 77
the entire new OneDrive, 18
The Files app over on the left-hand rail, 18
the Files tab, 45, 83
The Files tab, 7
the filter you can filter by name, 50
The final option, 37, 43, 68
The final point worth mentioning, 35
the idea behind Breakout Rooms, 81
the invited participants, 81
the latest features and tools in Microsoft, 2
the latest files in OneDrive, 125
the live captions are rolling., 21
the main Channel, 14
the main web area, 19
The Meet app is now available, 19
the meetings app, 18
the Microsoft ecosystem, 125
the More Options area, 61
the new OneDrive, 125
the new Teams, 18, 109
The next option is to raise your hand., 59
The one highlighted in purple, 6
the organizers and the co-organizers, 11
the original due date, 99
the pandemic, 6, 52, 60
the pandemic as remote work, 6
the participants will see live any changes, 70
the people needed for that project channel, 82
the platform, 1, 5
the portal, 6
the private teams, 23
the process of setting up virtual meetings and calls, 2
the project communication using posts, 82
the Scheduling Assistant, 63
the Share button, 69, 71, 76
the sidebar, 118
The Stream app, 19
the student icons, 107
the Team code area, 33

the team owners, 33
the team settings, 34
the team structure, 25
The third step is building project groups, 82
the title, date, time, and attendees., 81
the top corner, 60, 65, 75, 76
the UI in the new Teams, 14
the various apps and add-ons, 2
the web browser, 6
the website, 20
the whole umbrella kind of team., 82
There are a few ways that you can view your calendar, 13
there's nothing on your agenda, 13
think about different ways, 128
Think of channels as focused work streams, 82
third option, 26
third solution, 122
third-party apps, 10
This book is about the newest features and improvements, 2
thread of information, 109, 110
three main sections., 85
three options, 36
three other people, 39
three-dot menu, 17, 21, 104, 124, 125, 126
three-tier hierarchy keeps everything aligned, 82
thumbs up icon, 43
time limit for people, 79
time of recording, 110
time-saving assignment, 95
time-saving assignment updates, 95
TIPS AND TRICKS, 109
tips and tricks along with some basics, 109
To-do into one centralized area, 85
together communication, 82
top right-hand corner, 49, 52, 55, 56
Track status, 83
Training, 28
transcripts of the meeting, 54
trash can icon, 44
troubleshooting steps, 2, 123
TROUBLESHOOTING TEAMS, 119
trying out the activities, 1
trying to find meetings pretty quickly., 13

turn on live captions, 21
turning on automatic, 113
two different sections, 69
two types of teams, 23
type of channel, 34
type of digital activity, 108
types of channels, 26

U

uncheck spell check, 121
understand how to use Microsoft Teams in 2024,, 1
UNDERSTANDING TEAMS, 3
universal search, 11
unsave a message, 47
update in real time, 72
Updated Turn in celebrations, 105
upload a PDF document, 108
upload custom apps, 34
upload documents, 9
urgent messages, 69
use advanced features, 2
use emojis, 5
use external tools within Teams, 2
use Microsoft Teams effectively, 1
use Teams to join a meeting, 5
use the last option, 13
use the new view called Agenda, 13
use the search bar to find posts by keyword or name, 83
use Wi-Fi, 120
use your formatting, 73
user interfaces and features,, 20
users and Mac users, 6
Users can have one-on-one or group chats with colleagues, 5
using a free tool such as MS Teams, 84
using a mouse, 97
using different styles, 51
USING MICROSOFT PLANNER IN TEAMS, 85
USING MICROSOFT TEAMS, 82
USING MICROSOFT TEAMS FOR PROJECT MANAGEMENT, 82
using mobile data, 120
using tabs to integrate external tools, 83

using Teams channels, 109
using the Apps button, 10
Using the illustration, 86
Using the search bar, 115
Using the Upload button, 18
utilize templates, 73

V

version of Teams, 107, 122
very final point, 50
video and audio, 49, 52
video meeting, 55
video settings page, 56
videos, 20, 74, 117, 128
view by day, 13
view the recap, 18
virtual backgrounds, 5
virtual collaboration, 6
virtual collaboration have become increasingly prevalent., 6
Viva insights, 126
Viva Insights, 2

W

Warning for missing attachment, 103
We can create a basic team from scratch, 25
We can create brand-new documents, 9
We can create our own channels, 41
we can edit messages, 48
we can have conversations in, 41
We can just create a progress bar, 86
we have a **search bar**, 11
We have another channel for monthly reports, 6
we have our new team, 28
we have our profile picture, 47, 52
web version of OneDrive, 18
web-based version, 6
web-based version accessible, 6
we'll find the assignment, 96
we'll flip over to the student, 96
We're going to look at our app launcher, 6
We've got the calendar, 11
we've got three different notifications, 17

We've talked about teams, 52
whiteboard, 65, 72, 73, 74, 75, 76, 114
Whiteboard, 14, 72, 73, 75, 76, 113, 114
whole new window, 40
whole team, 34
Why did you feel successful recently, 101
Why do you need this book?, 1
wide range of third-party Integrations, 5
wide range of third-party Integrations and apps, 5
wondering about class materials, 18
Word, 5, 9, 14, 18, 40, 104, 125
work better with your team, 1
workbooks, 9
working problems, 120
working with conversations in Teams, 42
working within the Whiteboard app, 114
write a task for yourself, 110

Y

you also have access to some analytics for the channel., 35
you also have options to join Teams meetings, 66
You also have some settings for our rooms, 79
You can add a new task, 14
You can add the title for the meeting, 63
You can also add new members using the add member link, 32
You can also choose to turn off your camera, 56
You can also embed web applications, 83
You can also make an announcement, 78
You can also rename rooms, 79
You can also right-click to pin, 20
You can also use @ mentions in your conversations, 45
You can change your view, 59
you can chat with your colleagues, 42
you can choose for your live captions, 22
You can choose to mark all, 40
you can choose to view it based on the Work week, 13
you can control notifications, 53
you can copy the link, 41
You can create a task from a message, 110
you can create a Teams meeting, 67
You can create new rooms, 78

You can create your own whiteboards, 73
You can get one pinned into your channel, 128
You can get Teams to automatically identify you in the meeting captions and transcripts, 54
you can go random or choose one, 100
You can integrate external tools, 83
You can join the meeting, 66
You can mute that chat, 40
You can now use Teams to join a meeting or start a conversation., 5
You can reschedule, 125
you can see who has shared access, 15
you can start an audio call, 55
you can use your own image, 56
you could add an emoji, 43
You have a little Settings icon, 57
You have a meeting info area, 61
You have a Tags section, 34
You have an @mention section, 33
You have an Apps tab, 33
you want to create and from the list, 77
you want to have lots of people in your meeting., 77
you will learn about the latest features, 2
You will learn how to post and format messages, 2
You will learn how to set up Teams from various sources, 2
you'll be a pro at using Microsoft Teams, 1
you'll be able to drill in and see reading accuracy, 108
you'll get two options, 44
You'll see lots of entries, 16
you'll see your task planning Loop component, 14
your audio, 57
your chat messages, 51
your colleagues, 42, 76
your computer, 13, 44, 69, 122
your employer, 1
your entire screen, 68, 69
your fingertips, 85
your little camera toggle, 61
your microphone, 57, 62
your notifications, 40
your personal life, 128
your portfolio scales up, 82
your recommended videos, 20
your related projects, 82
your team for access, 83
your whiteboard, 73, 74, 114
YouTube, 117

www.ingramcontent.com/pod-product-compliance
Lightning Source LLC
Chambersburg PA
CBHW082331220526
45470CB00008B/2479